Henry W. Sang, Jr.
Fall 75

11/3

25-
(B)

Chinese Science

Explorations of an Ancient Tradition

Compiled in honor of the seventieth birthday of Joseph Needham, F.R.S., F.B.A.

M.I.T. East Asian Science Series
Nathan Sivin, general editor

Volume II

Chinese Science

Explorations of an Ancient Tradition

Edited by
Shigeru Nakayama and
Nathan Sivin

The MIT Press
Cambridge, Massachusetts, and London, England

This book was set in Monotype Baskerville,
printed on Warren's Old Style Offset
by The Maple Press Company
and bound by The Maple Press Company
in the United States of America.

Jacket illustration is a diagram showing the phases of the moon, from *Yuan t'ien t'u shuo* (illustrated explication of the heavens, 1819).

Library of Congress Cataloging in Publication Data
Main entry under title:

Chinese science.

(M.I.T. East Asian science series, v. 2)
"Compiled in honor of the seventieth birthday of Joseph Needham, F.R.S., F.B.A."
CONTENTS: The historian of science as ecumenical man: a meditation in the Shingon temple of Kongōsammai-in on Kōyasan, by J. Needham.—Joseph Needham and the science of China, by D. J. de Solla Price.—Joseph Needham, organic philosopher, by S. Nakayama. [etc.]
1. Science—History—China—Addresses, essays, lectures. I. Needham, Joseph, 1900– . II. Nakayama, Shigeru, 1928– , ed. III. Sivin, Nathan, ed. IV. Series: Massachusetts Institute of Technology.
M.I.T. East Asian science series, v.2.
Q 127. C5C555   509'.51   72–5577
ISBN 0–262–14011–X

Contents

The M.I.T. East Asian Science Series

One of the most interesting developments in historical scholarship over the past two decades has been a growing realization of the strength and importance of science and technology in ancient Asian culture. Joseph Needham's monumental exploratory survey, *Science and Civilisation in China,* has brought the Chinese tradition to the attention of educated people throughout the Occident. The level of our understanding is steadily deepening as new investigations are carried out in East Asia, Europe, and the United States.

The publication of general books and monographs in this field, because of its interdisciplinary character, presents special difficulties with which not every publisher is fully prepared to deal. The aim of the M.I.T. East Asian Science Series, under the general editorship of Nathan Sivin, is to identify and make available books which are based on original research in the Oriental sources and which combine the high methodological standards of Asian studies with those of technical history. This series will also bring special editorial and production skills to bear on the problems which arise when scientific equations and Chinese characters must appear in close proximity, and when ideas from both worlds of discourse are interwoven. Most books in the series will deal with science and technology before modern times in China and related Far Eastern cultures, but manuscripts concerned with contemporary scientific developments or with the survival and adaptation of traditional techniques in China, Japan, and their neighbors today will also be welcomed.

Volumes in the Series

Ulrich Libbrecht, *Chinese Mathematics in the Thirteenth Century: The Shu-shu chiu-chang of Ch'in Chiu-shao*

Shigeru Nakayama and Nathan Sivin (eds.), *Chinese Science: Explorations of an Ancient Tradition*

Manfred Porkert, *The Theoretical Foundations of Chinese Medicine: Systems of Correspondence*

Sang-woon Jeon, *Science and Technology in Korea: Traditional Instruments and Techniques*

Contributors and Areas of Study

William C. Cooper, M. D., Director of Clinical Research, Ross Laboratories, University of Ohio (clinical medicine)

A. C. Graham, Professor of Chinese, School of Oriental and African Studies, London University (Chinese philosophy, logic, grammar, and poetry)

Ho Peng Yoke (Ping-yü), Professor of Chinese Studies, University of Malaya, Kuala Lumpur (history of science in China)

Beda Lim, University Librarian, University of Malaya (library development)

Saburō Miyasita, Librarian, Takeda Chemical Industries, Osaka (early East Asian materia medica and medical bibliography)

Francis Morsingh, Professor of Phytochemistry, University of Malaya (alkaloid synthesis)

Shigeru Nakayama, Lecturer, College of General Education, Tokyo University (history of Far Eastern and Occidental science, modernization)

Derek J. de Solla Price, Avalon Professor of the History of Science, Yale University (physics, history of science and technology, science policy studies)

Nathan Sivin, Associate Professor of the History of Science, Massachusetts Institute of Technology (Chinese scientific thought)

Kiyosi Yabuuti, Professor Emeritus, Kyoto University; Professor, Ryūkoku University (history of exact sciences in China and Japan; transmission of scientific ideas and techniques)

Mitukuni Yosida, Assistant Professor, Research Institute for Humanistic Studies, Kyoto (history of science and technology in Japan and China)

Preface

This book is meant to help its readers learn about, and think critically about, the development of science in traditional China. Some readers will be drawn to it by the idea that humanity has evolved more than one tradition of natural science that deserves to be taken seriously. Some will be curious that a civilization with such deep aesthetic and spiritual roots should have had a strong scientific dimension. Others will wish to explore the possibility that by reconstructing and imaginatively adopting the viewpoint of such a different scientific tradition, they might become more critical in judging what aspects of the West's Scientific Revolution grew out of local pressures and prejudices rather than out of the inner necessities of science itself. It is not in our power to offer ready-made solutions, but we hope that everyone who brings his or her curiosity to this book will find at least something of use in it.

This volume falls naturally into two parts, which we believe are complementary. The first part provides the reader with perspectives on the work of Joseph Needham, whose monumental survey of the Chinese technical traditions, *Science and Civilisation in China*, is largely responsible for the growing awareness on the part of inquiring people everywhere that these traditions reached a high level, and that the birth of modern science and technology owes a great deal to them. Ponderous disquisitions on why China had to get along without anything that could be called scientific thought, and with only primitive technology, are much less common now that Needham has shown with massive documentation in volume after volume that their premises are based on garden-variety ignorance. His work is not, despite its large scale, a definitive summary, but a preliminary exploration, a first synthesis of three centuries of piecemeal research bound together with original discoveries and bold but largely tentative hypotheses of his own. The "Meditation" that precedes Chapter 1 of this volume is one of the clearest expressions I have seen of the motive forces and ideals behind his life's work—of which the historical study of Chinese science has been only one part. Chapters 1 through 3 are meant as aids to reading *Science and Civilisation in China* critically, to being aware of it not only as finished historical narrative but also as an act of creating both intelligible history and hypotheses about scientific change. Price's article provides basic information about the series, its author, its genesis, and evolution; Graham looks closely at Needham's central concern with the effect of social and economic factors upon the rate of scientific and technological change; and Nakayama demonstrates how greatly Needham's own intellectual encounter with Chinese science was shaped by the original approach to the philosophy of science that he worked out as a young biochemist.

In the rest of the book we have gathered a sample, as representative as possible, of the kinds of exploration now proceeding at the various frontiers of Chinese science. Our canvass has been restricted to scholars who are bringing high intellectual and critical standards to their study of the primary documents. Since we have cast our net in the direction of such a small school of fish, it is not surprising that the catch is small. We regret the lack of attention, for instance, to technology and to mathematics.[1] We particularly regret our lack of success in soliciting contributions from the many able specialists in China. Despite these limits we can claim that in a small compass most contemporary approaches to exploration of the Chinese technical traditions are represented. We have attained, perhaps for the first time in a Western publication, something like a balance between Occidental and Japanese scholarship. The reader can compare sharply focused philological analysis (Graham and Sivin) with synthesis that relies on a broad historic sense honed by intimate acquaintance with the sources (Yosida). One can weigh an approach that measures particular achievements by the yardstick of modern science (Yabuuti, Miyasita) against the attempt to reconstruct the inner unity of ancient science by restoring what seems to be superstition and what seems to be anticipation of a modern position to their unity and coherence in the ancient Chinese thinker's mind (Cooper and Sivin). One can consider ideas mainly in relation to other ideas (Yosida) or see them as conditioned by (and conditioning) the society in

---

[1] The latter omission is perhaps redeemed by the first volume in this series.

which the men who thought them lived (Graham). One can leave the plane of historical interpretation and look at the documents themselves (Ho, Lim, and Morsingh). Finally one can define the direction of one's own exploration with the aid of the critical bibliography that ends the book.

In the remainder of this preface I shall try to formulate some rather fundamental ideas about what traditional Chinese science was, about fruitful approaches to comprehending it, and about what influence it is likely to have, once we have learned to take it seriously, on thinking about science as a worldwide phenomenon with many local variants. This last point seems obvious enough, but its implications are very different from those of considering the Occidental tradition as a distinct entity that alone had the potential for evolving modern science, and thinking of other traditions as merely flawed imitations that remained fixed at a primitive stage. I shall argue that, since the theoretical and practical approaches to nature seem in traditional societies everywhere to have formed a unity with the social, political, and spiritual aspects of life, the reader can enrich his understanding of the latter to the extent that he is aware of the former. Finally I shall attempt to summarize the contributions of Chapters 4 through 9 of this book, and to explain briefly its genesis.

The Sciences of Traditional China

There are two obvious ways of defining traditional Chinese science. One takes as its subject matter all systematic abstract thought about nature.

"Abstract" may have the sense not only of defining concepts on a more general plane than that of concrete sensual experience, but also that of seeking objective driving forces of change within nature itself rather than, like religion and magic, looking for explanations in terms of conscious will or emotion. The scientist takes as one of his basic rules of reasoning that, other things being equal, the simpler of two hypotheses is more likely to be worth taking seriously. As Galileo learned at great cost, this notion can become highly offensive to the institutional protectors of revealed religion, who see it quite accurately as setting limits upon Providence.

The second definition is apparently much more straight-forward. It takes as scientific any idea, discovery, or method that also plays a role in modern science. Modern science is necessarily defined as the state of science today, since some aspects of last year's science are already obsolete and we are not at all certain what will still be valid next year. Ideas, meth-ods, or discoveries that played a demonstrable role in the development of today's science but fell by the wayside tend in this view to be classified as protoscience or mistaken science. The difference between the two need not concern us here.

The second definition seems at first glance to differ from the first only in being somewhat narrower, but actually their re-lation is nowhere near so simple. What do we mean when we talk about an anticipation of modern science? When we claim that the ancient Chinese anticipated modern scientists in pre-paring steroid hormones and protein hormones by processing large quantities of urine, or realized the importance of equa-torially mounted astronomical instruments and daily obser-vations long before Tycho Brahe introduced these innovations to Europe, we do not mean that the Chinese were doing exactly the same thing in the same way as scientists today for the same ends—no more so, in fact, than we mean Tycho was doing the same thing in the same way for the same ends as the twentieth-century astronomer. We are simply making close and at best compelling analogies. I doubt that there is a major publisher in the United States or Western Europe today who does not receive at least once a year a manuscript purporting to dem-onstrate in minute detail that the essentials of relativistic physics are to be found in the mystical epigrams of the *Lao tzu* (the book known in English as The Way and Its Power) or the cosmic commentaries of the Book of Changes. Here it is im-mediately evident that the only common ground is an analogy, and a weak one at that. Since the simple notion of relativity of space and time goes back beyond Einstein practically to the beginning of thought about nature in the West as well as the

East, such books seldom reach print.[2] The point of view that identifies ancient science simply as what preceded modern science and has survived in it is based on analogies that often become less solid and less pertinent the closer we look at them.

These two ways of defining early science even imply very different modes of exploration. Every historian of science uses the second (which I shall call the positivistic definition) to locate and identify anticipations of modern knowledge, either because he simply wants to document the time and place of a discovery, or because he wants to pick out of the immensity of recorded knowledge an event or idea the surroundings of which are likely to repay close exploration. One cannot, after all, study everything. It is natural to begin by trying to find patterns similar to those already known and understood, at least until one has probed the depths of the new structure and sensed what holds it together.

Finding anticipations is not merely a matter of passive acts of recognition, but rather, as I have suggested, a matter of actively constructing analogies. Democritus' speculations used to be studied as a precursor of modern atomic theory, until finally historians of science noticed that the unsplittable little solid Democritean atoms bouncing off each other in the void had very little in common with the present-day conception of the atom. Now they are likened somewhat more plausibly to the endlessly colliding molecules of thermodynamics.

If, on the other hand, we define early science simply as everything that was thought systematically, abstractly, and objectively about nature, our exploration takes the form of reconstructing what ideas of that sort developed, how they were related, how their correspondence to reality changed, and which of their

[2] Of course some such extravagances do reach print. Since writing the above words I have noticed on pp. 182–183 of Georges Beau, *La médecine chinoise* (La Rayon de la Science, Paris, 1965) an announcement—based on a misinterpretation of a mistranslation of a ludicrously misdated text—that an advisor of the Yellow Emperor forty-five centuries ago anticipated Einstein in revealing the identity of energy and matter, asserted the existence of immense forces situated in space, and knew of the planets Neptune and Pluto. The idea that many key concepts of modern science are latent in the *I ching* is developed in detail in Z. D. Sung, *The Symbols of Yi King or the Symbols of the Chinese Logic of Changes* (Shanghai: The China Modern Education Co., 1934).

possible consequences actually emerged in the course of time. Only after we have reconstructed these integral systems of ideas, I submit, can we be reasonably sure that in comparing the strengths of one scientific tradition with another we are comparing commensurable entities, rather than offering analogies that look compelling only so long as what we are analogizing is kept separate from its original context.

The two definitions lead, for example, to very different divisions of the fields of science in ancient China. The positivist whose main concern is early traces of modern science naturally classifies what he has found according to today's rubrics. Simple analogy makes Chang Heng, the inventor of the world's first known earthquake-indicating machine in the first century B.C., a seismologist. The early evaporators of urine become, by a subtly more subtle reasoning, early biochemists. At the extreme of this approach, as I have already noted, Lao-tzu the legendary mystic becomes a relativistic physicist.

The less forced labels of this kind perform several valuable educational functions. They help to dissuade people whose real interest is modern science from ignoring ancient science completely. They remind the average educated reader, who has all he can do to cope with the present and who finds most history good for nothing except entertainment, that modern physics did not spring full-blown from the collective brow of a few Nobel Laureates, and that it is not a solid structure of finished truth but a moment's transitory understanding in an apparently endless quest for comprehension. They may even help to convince him that modern science has a long way to go before it completely sheds the European parochialisms which it incorporated during the Scientific Revolution. They have also served a useful purpose in reducing the psychic dependence of contemporary Chinese scientists and engineers upon imported models of research and invention, encouraging them to consider carefully what approaches best fit their own circumstances.[3]

[3] The continuity of traditional scientific and technological themes and values in present-day China, and their role in the Cultural Revolution, has

Still there was no such coherent field within Chinese thought as biology or physics.[4] We must glean our anticipations of the modern science of life from far and wide, from sources that their authors would not have agreed had anything significant in common. Much of what early Chinese thinkers had to say about motion, heat, and light is just as scattered. The basic concepts that they used to explore physical phenomena are precisely yin-yang, the Five Phases (*wu hsing* 五行), the trigram and hexagram systems of the Book of Changes, and others that used to be invoked (even by twentieth-century Chinese thinkers) as chiefly responsible for the failure of Chinese to learn how to think scientifically.

But to place the responsibility there is to commit one of the most elementary fallacies of historical explanation, namely to present a description of what the world was like before X as though it were an explanation of why X happened so late or failed to happen at all. Since the concepts I have mentioned *were* the vocabulary of early scientific thought, it is as misguided to call them an impediment to modern science as to consider walking an impediment to the invention of the automobile. But that fallacy is a very tempting way out of difficulties into which we inevitably fall if we ignore the limitations of positivistic analogies and take them for full explanations. Needham has given prominence to anticipations of modern science, and drawn upon it for his rubrics, because his aim is to demonstrate the worldwide character of the development of science and technology. But he has consistently probed deeper, and has made sense of many Chinese concepts and shown their connections for the first time. Nevertheless all indications are that in the "Chinese science boom" now gradually taking shape the need for this kind of balance will all too often be ignored. Most of the recent tomes on Chinese medicine are, if occasionally better informed

---

been suggested in a most illuminating book by Yamada Keiji 山田慶兒, *Mirai e no toi: Chūgoku no kokoromi* 未來への問, 中國の試み (Inquiry into the future: The Chinese experiment; Tokyo: Chikuma Shobō, 1968).

[4] I have developed this point in *Chinese Alchemy: Preliminary Studies* (Harvard Monographs in the History of Science, I; Cambridge, Mass.: Harvard University Press, 1968), pp. 7–8.

than their predecessors, no more critical, disciplined, or conscious of the vast gaps between the elementary assumptions and definitions of the Oriental science and those of modern anatomy, physiology, and pathology.

If we use the first definition of science as systematic abstract thought—which might be called "anthropological," since it lets the Chinese theoretical encounter with nature define its own boundaries—the structure of knowledge looks very different indeed from that of modern science. It becomes as individual, in fact, as the map of scientific thought in ancient Greece or medieval Europe.

Although every culture must experience much the same physical world, each breaks it up into manageable segments in very distinct ways. To make a long story short, at the most general level of science (natural philosophy, it used to be called) certain basic concepts became established because of their very general usefulness in making nature comprehensible. In Europe after Aristotle's time among the most important of these notions were the Four Elements of Empedocles and the qualitative idea of a proper place that was part of the definition of each thing. In ancient China the most common tools of abstract thought were the yin-yang and Five Phases concepts, implying as they did a dynamic harmony compounded out of the cyclical alternation of complementary energies. Today scientists use a much wider range of well-defined concepts, embracing space, time, mass, energy, and information.

Thus the fields of science in a given culture are determined by the application of these general concepts, suitably refined, reinterpreted if necessary, and supplemented by more special concepts, to various fields of experience, demarked as the culture chooses for intrinsic and extrinsic reasons to demark them.[5]

It is natural to expect that, given roughly the same body and much the same range of malfunctions likely to befall it, every

---

[5] One remembers the puzzlement of a great European historian of China when he discovered that the Five Phases system in medicine did not quite correspond to that in astronomy. The difference is quite comprehensible in terms of the specialization of concepts.

culture will draw the boundaries of medicine more or less similarly. Actually this is far from true. One can find no consensus
even among modern physicians in the wealthy countries as to
how seriously emotional factors must be weighed when treating
somatic disease, and vice versa.[6] The student of intellectual
issues can distinguish various schools of thought, and the anthropologist can locate conventional views among those who
prefer not to think too deeply about these questions, although
even the clichés have changed drastically over the past century.
But the decisive mutual interaction of emotional and somatic
states has been one of the most constant doctrines of Chinese
medicine. In fact we find intricately developed theories in areas
that lie almost completely outside the intellectual horizons of
the modern doctor. There was, for instance, a strict correlation
of variations in the body's energetic functions with the cosmic
cycles of the day and the yearly seasons. Beginning in the T'ang
it was further elaborated in the *yun ch'i* 運氣 theory to take into
consideration in diagnosis the effects of unseasonable variations
in climate and weather.[7]

Then what fields of science did the Chinese themselves organize in the course of conceptualizing the phenomenal world?
I would propose more or less the following list of major disciplines:

1. Medicine (*i* 醫), which included theoretical studies of
health and disease (*i ching* 醫經, and so on), therapeutics (*i
fang* 醫方, and so on), macrobiotics (*yang sheng* 養生) or the
theory and practice of longevity techniques, sexual hygiene
(*fang chung* 房中, on the whole an aspect of macrobiotics),
pharmacognosy (*pen-ts'ao* 本草), and veterinary medicine (*shou
i* 獸醫). Pharmacognosy, the study of materia medica, incorporated a large part of early knowledge of natural history as
well as approaches to biological classification; Needham and
his collaborator Lu Gwei-djen generally refer to it as "pharmaceutical natural history." Acupuncture, which has recently

[6] See Chapter 8.
[7] See the third volume in this series, Manfred Porkert, *The Theoretical
Foundations of Chinese Medicine.*

been seized upon by the mass media in their endless quest for novel misinformation, was (and still is) on the whole only a minor branch of therapeutics.

Many of the proponents of Chinese medicine have sought to raise its status by portraying it as a purely empirical art or as a purely rational theoretical science—or both, despite the contradiction between an inductive structure, which grows out of clinical experience, and a deductive one, which depends upon the interrelations of ideas. To measure the strengths and weaknesses of Chinese medicine by the criteria of the completely different system that developed in the West gives an extremely distorted view of what it was about—as distorted a view as if a Chinese doctor were to praise European medicine only for those conceptions and methods that are also found in his own tradition. We need to make valid comparisons, but they must be derived from an objective and systematic description of Chinese medicine based on broad and deep acquaintance with it.[8] So far we can only begin to glimpse the outlines of a remarkably abstract and comprehensive system of concepts describing the relations of the body, the mind, the immediate physical surroundings of the body, and the cosmos. Chinese doctors commanded a great diversity of therapies and techniques, which they generally used in combination. Their classical tradition seems to have been more open to contributions from folk medicine than the written tradition of pre-Renaissance Europe (see Chapter 8). But every cure had to be integrated into the theoretical structure, in principle at least, before it could be

---

[8] Although no such general description exists for ancient medicine, traditional medicine as it was understood and practiced just before the Cultural Revolution is extremely well laid out in a volume compiled by the Nanking College of Chinese Medicine, *Chung-i-hsueh kai lun* 中醫學概論(An outline of Chinese medicine; Peking: People's Hygiene Press, 1958). The collapse of traditional values over the past century has made contemporary practice an utterly unreliable guide to the mainstream of early medicine, but critical study of this book, with its limitations kept in mind, is possibly the best available preparation for historical study of the documents. It avoids much more successfully than most publications from Taiwan, Hong Kong, and Europe the temptation to make offhand and ludicrous analogies with modern scientific medicine.

passed down in the literature. Drugs were assigned to functional categories that corresponded to the Five Phases system. Their combined use in prescriptions, their dosage, and the time and conditions of their administration were often either determined or accounted for by theoretical considerations. The effectiveness of curative agents was explained in terms of action upon the various functional systems of the body. Although in the larger and more pragmatically oriented therapeutic compendia these considerations were often not spelled out, every educated physician knew how to apply them. They were, in fact, his normal form of discourse. The intimate but extremely complex accomodation between theory and practice seems in the light of our very limited knowledge to have had very little in common with that of modern medicine.

I would respectfully suggest that any modern physician who finds it easy to think of traditional Chinese medicine as merely a local variant of his own inductive science has not been looking at the Art of the Yellow Emperor closely enough. A Paracelsian physician of the seventeenth century might perhaps have been more entitled to the shock of recognition.

2. Alchemy (*fu-lien* 伏鍊, and so on), the science of immortality, which overlapped greatly in practice with medical macrobiotics. Immortality was thought of, in fact, as the highest kind of health. The two major divisions were "external alchemy" (*wai tan* 外丹) and "internal alchemy" (*nei tan* 內丹). In the former, immortality drugs were prepared by techniques largely based on the natural processes by which minerals and metals were believed to mature within the earth, but in the laboratory they were carried out on a telescoped scale of time. A year of cyclical treatment by the alchemist might correspond to a cosmic cycle of 4320 years. In internal alchemy, by a different sort of analogy the interior of the adept's body became the laboratory and the "cyclical maturation of the elixir" was carried out by meditation, concentration, breath control, or sexual disciplines.

External alchemy, as I have shown elsewhere, was on the whole, despite the employment of a great deal of miscellaneous chemical knowledge, antecedent to physics rather than to chemistry. The adept was not directly concerned with exploring chemical reactions, but with designing laboratory (or psycho-physiological) models of cosmic process.[9] The laboratory models were partly chemical and partly physical in character; the ancient practitioner was not concerned with that particular distinction of ours.

3. Astrology (*t'ien-wen* 天文), in which anomalous celestial and meteorological phenomena were observed and interpreted in order to detect defects in the political order. This science was based on a close correspondence between the cosmic and political realms. In the "field allocation" (*fen yeh* 分野) theory of the second century B.C., this was an actual mapping of sections of the sky upon political divisions of the civilized world. In astrology the Emperor was the mediator between the orders of nature and humanity. Like a vibrating dipole, the ideal monarch drew his charisma from the eternal order of the cosmos, radiating it in turn to inspire virtue—defined implicitly as values oriented toward hierarchical order—in society. Because of his centrality, the harmony of above and below depended critically upon the Emperor's ability to maintain his ritual and moral fitness. Omens in the sky were thus an early warning that his responsibility as Son of Heaven needed to be taken more seriously.

4. Geomancy (*feng-shui* 風水), the science of "wind and water," which determines the auspicious placement of houses and tombs with respect to features of the landscape. While alchemy uses the yin-yang, Five Phases, and other concepts mainly to study the temporal relations involved in maturing the Great Work, geomancy adapts them predominantly to topological configurations. Geomancy has been shrugged off as mere superstition, but perhaps more to the point is the rationale it provides for expressing the status or wealth of a family in terms of control of its physical environment in life and death.

[9] See my contribution, "The Theoretical Background of Chinese Alchemy," to Joseph Needham, *Science and Civilisation in China*, V (Cambridge, England: at the University Press, in press).

Nevertheless geomancy, which has no Occidental counterpart, is much more than an arbitrary excuse for demonstrating social clout. Geographers have found it intrinsically interesting as the world's only time-proven theoretical approach to the aesthetics of land use. In other words, it succeeds consistently in producing sites that, in addition to their supramundane virtues, are beautiful. No serious historical study or conceptual analysis of geomancy has yet been published in any modern language, including Chinese.

5. Physical studies (*wu li* 物理). The shape of this composite field has more in common with that of Greek or Islamic natural philosophy than with that of post-Newtonian physics. In general this is the area in which fundamental concepts were adapted and applied to explain particular physical phenomena, as well as chemical, biological, and psychological phenomena that were not distinguished or were thought to be closely related. We can distinguish three overlapping exploratory approaches, which at various times and in various circumstances served much the same function in Chinese thought as elementary physics today—as well as other functions with which modern physics does not concern itself. Their free use of numerology should not obscure their essentially qualitative character.

The two major early approaches to the theoretical principles behind the events of nature might be called "mutation studies" and "resonance studies." The first uses the conceptual apparatus of the commentaries to the Book of Changes, and is in general as concerned with the social and political spheres as with nature. The second, which unlike the others has been studied from the viewpoint of the history of science, has its own literature, going back to the second century B.C. The resonance (*hsiang lei* 相類) treatises elaborate the notion that physical interactions are prompted by or controlled by categorical associations and correspondences, set out in terms of yin-yang and the Five Phases.[10] The third approach is "correspondence

[10] Ho Ping-yü (Peng Yoke) and Joseph Needham, "Theories of Categories in Early Mediaeval Chinese Alchemy," *Journal of the Warburg and Courtauld Institutes,* 1959, *22*: 173–210; Yamada Keiji, "Butsurui sōkan shi no seiritsu

studies" (*ko chih* 格致, *ko wu* 格物), which attained prominence much later as part of the program of Chu Hsi 朱熹 (1130–1200) and other Neo-Confucian philosophers.[11] This tendency brought to bear on interesting natural phenomena not only the more sophisticated concepts of the Neo-Confucians but their concern for the didactic applications of their insights, for they were committed to the integration of nature, society, and the individual psyche. The term *wu li* (literally, "the pattern-principles of the phenomena") was ultimately redefined to become the standard equivalent in modern Chinese for "physics."

In addition to these qualitative sciences there were others concerned with number and its applications:

6. Mathematics (*suan* 算), which was on the whole numerical and algebraic in its approach rather than geometric, and oriented toward practical application rather than toward exploration of the properties of number and measure for their own sake. The search for the deeper meaning and implication of number seems to have remained within the province of what we would call numerology.

7. Mathematical harmonics (*lü* 律 or *lü lü* 律呂, Needham's "acoustics") was perhaps the field in which mathematics and numerology were applied in closest combination. It arose from discoveries about the simple numerical relations between sound intervals, which had also suggested to the Pythagoreans that the basis of regularity in nature was numerical. In China the relations explored were mainly those of the dimensions of resonant pipes. The importanc of emusic in ceremonial made harmonics part of the intellectual trappings of imperial charisma. This tied it to other kinds of ritually oriented charisma, guaranteed

---

物類相感志の成立"(The organization of the *Wu lei hsiang kan chih*), *Seikatsu bunka kenkyū* 生活文化研究, 1965, *13*: 305–320. It will be shown in *Science and Civilisation in China*, V that many of the sources of the Ho and Needham article were not concerned with laboratory alchemy but with physiological and psychic "inner alchemy."

11 *Science and Civilisation in China*, I and II, s. v.; D. C. Lau, "A Note on Ke Wu 格物," *Bulletin of the School of Oriental and African Studies*, London University, 1967, *30*: 353–357.

it sponsorship, protected its study, and petrified it institutionally. The very special nature of its connections with other kinds of activity tied to dynastic legitimacy, particularly mathematical astronomy and metrology, needs to be studied further. Attempts were made, for instance, to use the standard pipes as basic measures of length, capacity, and (indirectly) weight.

8. Mathematical astronomy (*li* 曆 or *li fa* 曆法) was, especially in early times, thought of as closely related to harmonics. In some of the dynastic histories the state of the two fields is surveyed in a single Treatise on Harmonics and Calendrical Astronomy (*lü li chih* 律曆志). Astrology and astronomy, on the other hand, were treated separately except in the first of the histories (*Shih chi* 史記, ca. 90 B.C.). Astronomy and harmonics both dealt with phenomena governed by simple but constant numerical relations. It was usual in early astronomy to rationalize and account metaphysically for the observed periodic constants by "deriving" them through numerology from the categories of the Book of Changes.[12] It is perhaps most useful to think of mathematical astronomy as aimed at making celestial phenomena predictable and thus removing them from the realm of astrology, which interpreted the ominous significance of unpredictable phenomena. The ties of both disciplines to ritual thought and institutions, which depend in their essence upon precedent, often tended to blur the boundary by retaining astrological significance for phenomena that astronomers had learned to predict. The sky's eternal changes, especially the fine variations underlying the gross regularities of the solar and lunar motions, prompted the most sophisticated technical developments of which Chinese mathematics was to prove capable.

This list of sciences will no doubt soon be revised as the result of deeper study. The alert reader will already be aware that its distinctions are far from clear-cut. Geomancy and alchemy, for instance, also dealt from their own viewpoints with the subject

[12] See Chapter 5 of this book, and for an example of numerological derivation, my "Cosmos and Computation in Early Chinese Mathematical Astronomy," *T'oung Pao* (Leiden), 1969, *55*: 8–9.

matter of what I have called physical studies, and what quantification physical thought achieved was in connection with
harmonics. In the West before recent times too there was no
need for sharp divisions between fields of science. Demarcations
depended largely on ideas about the hierarchic character of
knowledge, until finally the social specialization of research and
its applications became the overriding factor. Chinese did not
draw epistemological boundaries between the modes of qualitative exploration I have described. The fundamental distinction instead seems to have been that between fields accepted by
conventional people as orthodox and those set apart as hetero-
dox or marginal. This judgment depended on the tolerance of
the person doing the evaluating and the breadth of his conception of Confucian relevance; the point was, of course, what
sorts of conduct were considered appropriate to various social
roles. There was in many periods a consensus among men of
letters that knowledge of the cosmos was pertinent to the
betterment of society and thus a suitable concern for the gentleman, and at certain times a consensus that it was not. Nevertheless it is always necessary to question the precise social situation and motives that underlay any particular assertion about
what should or should not be studied. Generalizations such as
"Confucians believed" and "Taoists believed" perpetuate assumptions about effective social groupings and thus foreclose
the search for the real loci of belief. Such generalizations
amount to a self-imposed impediment to the serious study of the
sociology of science in China.

Astrology, astronomy, and harmonics were on the whole not
only orthodox but the special concern of the imperial bureaucracy. At times study of the sky was severely forbidden outside
the civil service, for fear of putting a powerful tool of dynastic
legitimacy (and thus of antidynastic rebellion) into private
hands. Medicine straddled the borderline. The numerical
majority of therapists must have occupied social positions close
to those of the commoners they treated by rough and ready
methods. They were often derided as quacks by educated physicians for whom deep draughts of Confucian philosophy and

Buddhist ethics came to be as essential as study of the fivefold and sixfold energetic systems of the body. Geomancy and alchemy were, broadly speaking, not respectable occupations for the Confucian gentleman, although a man who was already respectable might have a good deal of freedom to dabble in them (or in the formal cults of religious Taoism, for that matter) if he chose to exert it. Of course we must depend on written records produced by people who belonged to the educated minority (or minorities). Even from this skewed sample we can see that many of the practitioners who played an important part in the conceptual development of these two fields were marginal members of the élite. Alchemy and medical therapy in particular were associated with the Taoist ideology, which as an alternative to conventional attitudes evolved its own standards of behavior, increasing the diversity of roles and styles in Chinese society.

Chinese Science and Contemporary Issues

This crude topography of the Chinese sciences can at least serve to remind us of a truth that we easily forget when we focus our attention narrowly on anticipations of modern chemistry, physics, biology, and so on. When Needham speaks of internal alchemy, the art of perfecting the Elixir within the adept's body, as a "proto-biochemistry,"[13] he has already scrupulously depicted its methods as a variety of breath manipulations, sexual disciplines, and psychic meditations; but it will be interesting to observe how many writers in the next decade or so describe the initiates as if they would be more at home in a university laboratory than in a mountain abbey. There is no surer road to triviality than to forget that, although many Chinese concepts and attitudes can be found in science today—and, very likely, others in science tomorrow—the disciplines in which they were originally embedded were drastically different from our own in aim, approach, and organization.

---

[13] *Science and Civilisation in China,* V.

The shape of this ensemble of sciences reflected the changing configurations of Chinese society and the attitudes of various social groups in various periods toward nature (ranging from the wish to exploit it without hindrance to the mystical longing for union with it). We can recognize that astrology and mathematical astronomy developed as vessels for conserving not only attitudes about the sky and humanity's relation to it, but also cosmological aspects of the ideology of monarchy. It has been nearly forty years since Wolfram Eberhard began looking for ways in which astrology was used reciprocally to influence and change this ideology. We can see Chinese medical theory both as a creative act of understanding in terms of generally accepted concepts, and as a force for making that understanding universal within the groups brought up with those concepts. Classical medicine provided the educated civil servant of late imperial China with ideas of health and disease consonant with the rest of his world view, just as folk medicine provided his uneducated wife and servants with ideas continuous with their own perspectives. This split is no odder than that between an engineer today who enjoys working on any problem for its own sake and his son or daughter who sees technology on balance as a race toward ecological suicide.

In the close Chinese fit between science, everyday experience, and the norms of social subgroups, we see great likenesses to what European science was before the Scientific Revolution, and to what the traditional science of Islam largely remains today. Some students of Chinese culture depict its philosophy of nature as unique because it was moral, because theories were so often formulated in such a way that they had repercussions for social order. It is quite true that today we do not consider nature an ethical order, nor human society a little cosmos. But it is most unusual to find a people before modern times who did not find principles of human conduct and organization in nature, either immanent or put there by divinities. Hebrew speculation, with its gaze concentrated on godhead, is one of the very few exceptions. We can still see the continuum of nature, society, and the individual in pre-Renaissance Europe, but the link was cut through, gradually and with unabating resistance, by what C. C. Gillispie has called "the edge of objectivity." Modern science has nothing to contribute to moral reflection except predictions of consequences. Society at large retains little confidence even in the objectivity of these predictions, since in any issue of public moment it is almost inevitable that both sides will produce duly accredited scientists to predict what are represented as diametrically opposite outcomes.

Modern science defines its concepts out of the mathematical abstraction of measurable events. Pythagoras would have agreed that here lies the true simplicity. For the purposes of contemporary physical or biochemical theory, the rich sensual world of everyday experience is too complex to be directly germane. Most of its texture is necessarily discarded in the process of paring down to rigorously definable entities and parameters that are not designed to appeal to everyday common sense. A great deal of what seems familiar and obvious in popular writing about science abruptly disappears when these ideas are translated back into science's own language.

As science becomes more able to deal with complex organisms and systems in their entirety, it will have less need to rely as it has in the past on the model of mechanistic physics, and will perhaps develop a texture closer to that of the individual's direct perception of nature. This does not mean that we can settle back and wait for the broken unity of knowledge to mend itself. The tremendous power of science to incorporate theoretically any phenomenon humans can learn to state mathematically, and to specify means for any end it can predict, has generated social consequences at a rate many times too fast to be integrated by any means that civilizations have yet evolved. Mass-consumption technological razzle-dazzle— election-year "massive offensives against cancer" and bureaucrats landing on the moon to perform their mechanical tasks and leave behind their plastic bags of urine—does not entirely obscure our view of theoretical science receding further into realms of abstraction as society continues to grope for something to replace the absolute moral sanctions Descartes expelled from nature three hundred fifty years ago. If there is to be a new unity of custom, belief, and knowledge, it will come from new modes of adaptation that we have so far failed to evolve. It is far from clear how they can evolve in societies where scientific illiteracy is the norm except among specialist functionaries. In early societies the élite was by definition the custodian of discourse about society and its relations to nature. Today the average American, European, or Japanese university graduate is unprepared to critically evaluate technical arguments on issues of armament or pollution, although he or she is expected to vote with these issues in mind.

One of the most fertile questions we can ask, in view of our contemporary crises, is exactly how science and other aspects of culture coexisted in unity earlier. Comparative history sets before us a range of examples. If we are willing to study them attentively on their own merits and to reconstruct their own inner connections and dynamics, we can provide ourselves with a series of priceless case studies from which to draw valid comparisons—not of static structures, but of the ceaseless interplay of social change affecting the directions of scientific thought, and scientific change calling forth the adaptive powers of societies. There is nothing new about the idea that comparison is the most fruitful approach to learning which aspects of science are universal, and which are merely local peculiarities. But what I am suggesting here is that everyone who is concerned about the central problems of our time has a stake in the deeper understanding of science in ancient China.

The Contents of This Book
I do not mean to leave the reader with the impression that these concerns permeate the papers in this book. I have merely sketched out my own view of some general implications of comparative studies of the world's scientific traditions. It is too soon to exhibit in one volume the explicit efforts of a number of students of history to establish a pattern for such work. Progress toward that end has hardly begun in relation to Chi-

nese science, the best documented of the major traditions. This volume as a whole is meant not to demonstrate this nascent approach, merely to encourage its emergence.

My remarks are, however, meant to suggest why my coeditor and I have assembled this book as a tribute to Joseph Needham, who has always portrayed modern science as a river in which the waters of many streams are inextricably mingled. In the course of broadening and deepening our integral understanding of traditional Chinese culture, practically every paragraph Needham has written has been designed to be world history, and to urge upon his readers a more humane perception of the future. Serious and widespread comparative study of the history of science would have been almost impracticable before the appearance of his work, but it is now inevitable.

Although each of the papers in this book is a contribution to our more accurate understanding of Chinese science, taken together they illustrate a wide range of approaches, methods, and aims. I shall briefly characterize each article to make this range explicit for the convenience of those exploring the field for the first time, as well as to lighten the load of those of my colleagues who can no longer find the time to read books before they review them. I shall take the papers up in the order of their appearance, which is arbitrary but very roughly parallels the organization of the Bibliography.

Mitukuni Yosida is one of the very few Japanese scholars of Chinese science who has shown much interest in a broad and interpretive approach to early natural philosophy. His reflections on the inseparability of nature and man in the Chinese universal order, and on the concepts used to express this order, range freely over Neo-Confucianism and Taoism, esoteric practices and landscape painting. Even readers who may disagree with some point of detail will find in Yosida's paper a synthetic strength that reflects that of the old historians of China and Japan. An acquaintance with their work is a great help in getting over one of the prevalent weaknesses of European and American historical writing, the tendency to formulate questions of the kind "Was *A*, *B*, or *C* the cause of *X*?"

and thus to overlook the real problem, the balance and interplay of *A*, *B*, and *C* that brought *X* about.

Kiyosi Yabuuti and the members of his seminar at the
Research Institute for Humanistic Studies, Kyoto University,
have explored the length, breadth, and depth of the Chinese
tradition. The four volumes of their research reports provide a
chronological and topical survey that complements *Science and
Civilisation in China* and is no less indispensable to the serious
student.[14] Many of Yabuuti's own monographs have been
devoted to reconstituting and explaining the methods and
rationales of the exact sciences, especially mathematical astronomy. His paper is a clear and straightforward exposition of
the special characteristics of astrology and astronomy in China,
as usual concerned as much with their social functions and
organization as with their technical basis. He suggests that
political unrest did not generally inhibit scientific activity, and
that China's geopolitical setting limited its openness to foreign
influence, whereas that of Europe allowed scientific change to
proceed in some countries even while attempts to retard it
were temporarily succeeding in others.

Graham and Sivin deal with a problem that has been vexed
for centuries, the precise meaning of the eight propositions in
the Mohist Canon (ca. 300 B.C.) that deal with shadow and
image formation. Because the texts of these propositions are
corrupt and use the technical terminology of a school that died
out in the third century B.C. or so—cutting short a brief but
remarkable development of scientific thought—there have been

---

[14] *Tōhō gakuhō* 東方學報 (Kyoto), 1959, no. 30, special issue on ancient
science and technology; Kiyosi Yabuuti (ed.), *Chūgoku chūsei kagaku gijutsushi no kenkyū* 中國中世科學技術史の研究 (Studies in the history of medieval
Chinese science and technology; Tokyo: Kadokawa shoten, 1963); Kiyosi
Yabuuti (ed.), *Sō-Gen jidai no kagaku gijutsushi* 宋元時代の科學技術史 (A
history of science and technology in the Sung and Yuan periods; Kyoto:
Research Institute for Humanistic Studies, 1967); Kiyosi Yabuuti and
Mitukuni Yosida (eds.), *Min-Shin jidai no kagaku gijutsushi* 明清時代の科學技
術史 (A history of science and technology in the Ming and Ch'ing periods;
idem, 1970). The tables of contents of the four volumes are translated in
Shigeru Nakayama, "Kyoto Group of the History of Science," *Japanese
Studies in the History of Science*, 1970, no. 9, pp. 1–4.

many divergent attempts to understand them, as well as other propositions on such subjects as space, time, and mechanics. Scholars of great talent have studied profoundly either their language or the possible physical situations that they reflect. This article attempts to be rigorous in both respects. By refusing to assume that the Chinese concepts are those of the modern textbook of optics, the authors have detected several crucial differences (see in particular the "Physical Interpretation" of Proposition B22). Even though many problems of interpretation remain, we can begin to think in a somewhat more informed way about alternative approaches to optical theory in ancient times.

The rhymed stanzas of Ho, Lim, and Morsingh are the mark of a *jeu d'esprit;* but this paper is also an important addition to our very small store of translated alchemical documents,[15] and one of especial interest from the viewpoint of the relations of alchemy with medicine. The translators' annotations help to uncover the rich data on pharmacognosy, natural history, and Taoist immortality lore imbedded in the text. They have also broken new ground in dating a text of this kind by analyzing its rhymes.

Cooper and Sivin have tried to dissect out of the documents of classical medicine cures that were originally adapted from the drastically different milieu of folk medicine. The ultimate aim is to throw light on the character of ancient folk beliefs and practice by studying the "transformations" by which these cures, many of them magical, were adapted to the abstract concepts that ancient intellectuals used to understand disease. The hypotheses of this paper are documented with translations of prescriptions using drugs derived from the human body. Some readers will wish to survey these hypotheses, presented in the introductory pages and the Conclusions, before proceeding to evaluate the detailed evidence in the body of the paper.

Miyasita's paper is our one example of a report documenting

---

[15] For a list of translations, see Sivin, *Chinese Alchemy: Preliminary Studies,* pp. 322–324.

a single discovery. This distinguished scholar of Chinese medical literature demonstrates that a mixture of datura alba, aconite, and other drugs was administered as an anesthetic for treating multiple fractures and removing arrowheads in the mid-fourteenth century. His citation of "what may be the earliest documented surgical procedure under general narcosis in the world" throws light not only on what Chinese physicians could do but also on what they did not do; the very narrow range of surgical procedures for which this anesthetic was specified was striking but characteristic. For all its brevity, Miyasita's paper is also a contribution to our knowledge of another very meagerly explored topic, the history of the use of consciousness-altering drugs in China.

Finally, the Introductory Bibliography is not meant as documentation of the articles. It is a selected and informatively annotated guide to basic books and articles on Chinese science in Western languages. It has been planned to serve those who want to begin exploring Chinese science and its background for their own purposes, whatever these may be. In order to make the Bibliography as useful as possible, considerable space is devoted to books on more general topics that are especially pertinent to the development of science and to bibliographies that will lead the reader to more specialized studies.

Let me add a short account of how this book took shape. In 1968 I was approached by the editors of *History and Theory*—at the initial suggestion, I believe, of Bruce Mazlish—who wished to publish a special number on Chinese science. I proceeded to invite contributions from every scholar known to me, accessible by mail, and able to write in English or Chinese, who is doing analytical and critical work of a high order in the primary sources of Chinese science. It became clear that, although too few of my colleagues were concerned in their work with the philosophy of history (the special concern of *History and Theory*) to justify a *Beiheft*, more than enough contributions would be available to form a volume useful to a much wider public. I therefore moved the project to the

M.I.T. Press, which had earlier been enthusiastic about it. We had in mind examples of a wide range of research styles and approaches. Because we felt that such a volume could also help to make the importance of Japanese work on Chinese science better known, Shigeru Nakayama was asked to be coeditor and to provide liaison with our Japanese contributors. The decision to dedicate this volume to Joseph Needham to mark his seventieth birthday had been taken long before, and now it became possible to answer the need that I had heard voiced so often for a compact introduction to Needham's work from several points of view. Both Price and Nakayama offered to revise and bring up to date their previously published articles. Chapter 2 will doubtless interest readers who have wondered about Japanese reactions to Needham's work and its Occidental reception. Graham also agreed to widen slightly the scope of his critique, previously commissioned by *Asia Major,* in view of our purposes. Since the resulting changes were made in time to be incorporated also in the *Asia Major* article, the two versions became closely similar.

We did not, at the same time, feel the need to perform the functions of a conventional *Festschrift,* since two of these were being compiled elsewhere.[16] I did not hesitate, therefore, to include more than one article in which my own name figured as coauthor. At this early stage, collaborative research is an extremely powerful tool in opening up new topics in more than a superficial way. The two papers are offered in the hope that they will encourage others to make use of it. Nor did it seem out of order to invite Needham to set down any reflections the contents of this volume might prompt.

It would be impossible to convey the acknowledgments of each of the contributors. I will merely note that editorial work

[16] One is being compiled by Robert Cohen and other philosophers of science at Boston University; the other, Mikuláš Teich and Robert Young (eds.), *Changing Perspectives in the History of Science. Essays in Honour of Joseph Needham,* is forthcoming from Heinemann Educational. The latter volume contains a bibliography of Needham's writings, excluding papers in biochemistry and experimental embryology and morphology.

on this volume proceeded not only at the editors' own institutions, but also burdened the generous hospitality of the Sinologisch Instituut, Leiden, and the Research Institute of Humanistic Studies, Kyoto. It was partly supported by grants from the National Science Foundation and the John Simon Guggenheim Memorial Foundation. Ruth Dubois and her colleagues at the Department of Humanities, M.I.T., deserve much of the credit for making large parts of the final manuscript readable. The whole final draft was read and the editorial work criticized by Robert Somers. We are pleased that the M.I.T. Press's professional team working on the Series has been joined by Muriel Moyle, whose indexes we have long admired.

N. Sivin

Kyoto, Japan
16 April 1972

Chinese Science

Explorations of an Ancient Tradition

The Historian of Science as Ecumenical Man: A Meditation in the Shingon Temple of Kongōsammai-in 金剛三昧院 on Kōyasan

One of the greatest needs of the world in our time is the growth and wide dissemination of a true historical perspective, for without it whole peoples can make the gravest misjudgments about each other. Since science and its applications dominate so much our present world, since men of every race and culture take so great a pride in man's understanding of Nature and control over her, it matters vitally to know how this modern science came into being. Was it purely a product of the genius of Europe, or did all civilizations bring their contributions to the common pool? A right historical perspective here is one of the most urgent necessities of our time.

As it happens, the history of science as it has grown up in the West has had one besetting deficiency, the tendency to trace only one line of development, that from the Greeks to Renaissance Europe. This is natural because what we may call distinctively modern science did in fact come into being only in Western Europe during the "scientific revolution" of the fifteenth and sixteenth centuries, culminating in the seventeenth. But this is indeed far from being the whole story, and to tell this part of it alone is to be deeply unjust to the other civilizations. And unjust here means both untrue and unfriendly, two cardinal sins which mankind cannot commit with impunity.

The conventional story is also natural because the mainsprings of the scientific revolution, whether of internal logic or of external social facilitation or inhibition, are of great significance to everybody. If it was the most momentous turning-point in all scientific history, naturally people of all civilizations are eager to know exactly how it happened. But that should not diminish the importance of understanding also what were the contributions from Asia which helped to make the scientific revolution possible, nor should it diminish our interest in those non-European discoveries and inventions which did not get incorporated into it, whether by failures of transmission or by lateness of time.

The conventional story is again natural because of the Euro-pocentric habit of mind which so many illustrious thinkers have long entertained, and have succeeded in mesmerizing others to accept. This sprang from many different roots—the antithesis drawn so sharply in ancient times between Greeks and barbar-ians; the claim of the people of Israel to be the chosen nation of God, a claim inherited in the uniqueness and exclusiveness of most versions of orthodox Christianity; and last but not least the immense material power in weapons and armaments which the combination of modern science and developing capitalism gave to Europeans. This assumption of domination came to be an almost ineradicable habit of mind, but uprooted in these times it has to be. "We are the people, and wisdom was born with us" was what the fools said in their hearts, but at the pres-ent day such foolishness is dangerous. What we want is not nature but grace, or rather that right mixture of both which Thomas of Aquino spoke of when he said that "grace does not abrogate nature, but supplies and corrects the deficiencies of nature." So without in any way diminishing the stature of Galileo we must learn to do more justice to Aryabhāta and Kuo Shou-Ching 郭守敬.

For more than a century scholars awake to the facts have struggled against the conventional view of the development of the sciences. For Indian studies a Colebrooke or a Filliozat, for Chinese science a Gaubil and a de Saussure, for the Turkish field an Adnan Adivar, have all labored to demonstrate the contributions of those cultures to the general fund. Even within the undeniable borders of the track of transmission which Mediterranean science took, it has been necessary for Neugebauer to uncover the remarkable achievements of the Babylonian and other pre-Greek Mesopotamian cultures, while the greatness of the Arabic centuries has been revealed by many outstanding scholars like Aldo Mieli, Kraus, Ruska, and Wiedemann, and in our own time Hossein Nasr continues this tradition. Abel Rey and the Egyptologists added knowl-edge of another great pre-Greek culture. And from far away Sylvanus Morley for the Amerindians and Cline for the Africans have added much of interest. Perhaps the patron saint of all these open-minded men could be the Syrian bishop Severus Sebokht of the sixth century, who after describing the Indian method of calculating with only nine signs, remarked that it would be good for all those who hymned continually the genius of the Greeks to realize that others also knew something.

What is wrong with the conventional story is that there were indeed others who knew something. And one cannot separate science from technology, pure science from applied science— the two intertwine inextricably. Modern science did not spring into existence fully grown, like Athene from the brain of Zeus; it drew from many origins, and though astronomy, mathematics, and some parts of physics were the spearhead, a long time was needed before the sciences of chemistry and biology came to their maturity. By a thousand capillary channels, like venules joining together to form a *vena cava magna,* influences came from all parts of the world. The Indian numerals and computations have just been mentioned, but there was a notable development of atomism there. China provided many things, the escapement of mechanical clocks without which there could be no accurate measurement of time, certainly in being by the eighth century; the basic method of interconversion of rotary and longitudinal motion, invented by the sixth, or such a fundamental device as the axial rudder for ships, demonstrable from the first century. Persia provided the windmill, al-Nafīs stated part at least of the theory of the circulation of the blood, and al-Razī laid the foundations of chemistry as distinct from alchemy, all by the end of the ninth century.

And there were the two great extras of the scientific revolution itself. While Euclidean deductive geometry and Ptolemaic planetary astronomy remain undeniable bases for the Galilean movement, there were two other offerings highly important for the development of modern science which were not European. Medieval Europe had done something in dynamics, but knew nothing of magnetism till the end of the twelfth century. All the work on that had been done in China, where people were worrying about the cause of declination before Europeans even knew of the existence of polarity. So also for chemists Paracelsus deserves to be thought of along with Galileo, and by his momentous statement that "the business of alchemy is not to make gold but to prepare medicines" he was bringing into Europe the age-old Chinese belief in elixirs of life, ultimate source of all medical chemistry and chemotherapy.

Sometimes discoveries could be independent and not incorporated in the general onward march until rediscovered later. Seki Takakazu's 關孝和 steps towards the infinitesimal calculus in Japan might be an instance of this. Sometimes things could happen independently in several countries. Thus it looks as if the telescope was invented simultaneously in Holland, Italy, England, and perhaps China. But even when discoveries were not genetically connected with the first rise of modern science we ought surely to celebrate and appreciate them nonetheless. The Maya zero might be one example, or the Chinese inventions of the seismograph or differential gearing (second century), or the higher reaches of pattern-welding and the combination of different kinds of iron and steel which were brought to such perfection by the Japanese.

Why, it may be asked, should we do this? Because, as the Confucian sages emphasized, mankind is one great family, and the scientific view of the world has clearly transcended all differences of race, color, and religious culture. By this of course I do not mean that narrow and dogmatic interpretation of the phrase adopted by nineteenth-century agnostics, but rather the understanding of Nature and control over her operations in the broadest sense. We are dealing here not only with the unjust because untrue but the unjust because unfriendly. Only by repairing and avoiding this injustice shall we have any chance of surviving as humankind, now that modern science has released the fearful intra-atomic energies of the sun, and brought to earth the possibilities of infinite self-destruction. In the good old words: if we do not hang together we shall all hang separately. To do this, no people must claim that wisdom was born with them.

The basic criticism of European triumphalism, to use a word much used in theology nowadays, is that it is lacking in love, the *agapē tou plēsiou* of the Gospels. Equality is the essential watchword, the recognition that every people and every culture has contributed to that which everyone now most cherishes, the scientific view of the world. Moreover, as universal education in science spreads, they will do so more and more. Already we entrust our lives to each other freely, flying with an Arab pilot, a Madagascan copilot, a New Zealand navigator, and an Ethiopian in charge of ground computer control. Let us have at least the generosity to salute all that their past cultures did toward the science and technology of the ecumenical world.

In the Kongōsammai-in, early in the morning, the dean and the four canons take their places before the image of Amida 阿彌陀佛, and begin their chanting, while the lay folk, one by one, offer incense. Westerners, I reflect, still suffer seriously from the Four Evil Minds, greed, anger, foolish infatuation, and fear. Of greed, built into the structure of capitalist civilization, not much need be said. Anger or irritation at ways not their own is often to be found; and is not the idea that Western ways are the only universal culture a foolish infatuation? Fear, unreasoning fear above all, the fear of the alien and the unknown that expresses itself in phrases like "hordes of gooks," is an evil thing that Westerners must learn to overcome.

The Buddhist faith is not my own, but the temple is numinous in the extreme, and to this a Christian cannot but respond, making his orisons in his own way. While the dean and canons chant the saving power of Shakyamuni and all Buddhas and Bodhisattvas, making memorial also of their own first great abbot on this mountain, Kōbō Daishi 弘法大師, I praise the glory of the Holy and Undivided Trinity, hoping that this too may be acceptable to the eternal Tao 道, the Tao beyond speech, that cannot be spoken about. But if I pray in my own way that does not mean that a Christian cannot meditate on the symbolism and significance of what is before his eyes, on all the truth that he can find in Buddhism. The Great Renunciation has meant, and will mean, many things beyond the touching

parting of Gautama from his family to go alone and wrestle
with the mystery of the universe beneath the tree of illumina-
tion. In this main hall, Amida promises rebirth in the Pure
Land of the West, whence entry into *nirvana* will be easier; but
for us Westerners would it not be good to begin with the renunciation of our most dreadful characteristic, spiritual pride?
Not of course only what William Law and other old theologians
meant, but the fixed idea that we are the people, and wisdom
was born with us.

Perhaps this is where the Marxists (somewhat to their sur-
prise, no doubt) contribute to a juster humility of Europeans.
For if it was largely external social and economic circumstances
that brought about the scientific revolution, then the intellectual capacities of Westerners may not be (as so many of
them have believed) that much superior to those of other
peoples. The sheer intellectual power of the giants of the
Newtonian tradition might blind many (as it has tended to do
in the past) to the achievements of other traditions, and perhaps
if the social and economic circumstances had been right in
other cultures, modern science would have arisen somewhere
else than where it actually did. But these are deep matters
into which much thought and research must yet be poured.

And now the canons have moved into the adjoining chapel
on the left, while the dean is making an address to the laity on
the evil of sex unaccompanied by love; one can hear them
chanting an auxiliary liturgy there. That smaller hall is not
decorated with such dazzling magnificence of candles and
lamps, altar paraphernalia for offerings, and gilt shrines for the
commemoration of the departed, as is the main temple. But
it contains, I know, an image of the Buddhist tutelary god of
love, Aizen Myō-ō 天弓愛染明王 (Vajra-rāga-vidyārāja—Lord
of the Magic of Loves and Melodies), looking very fierce,
with six arms and six hands. This deity is not unconnected, I
suspect, with Indian Kāma and with the Tantric goddess or
Vidyārājñī Tārā, so important in Chinese Taoist and Buddhist
history. One remembers that I-Hsing 一行, the greatest astronomer and mathematician of his age, the seventh and eighth-

century T'ang, was a Tantric monk—indeed, he is one of the eight Shingon patriarchs. For Tantrism, unlike primitive Buddhism, was world-affirming rather than world-denying; and for it material things were not in themselves evil, because impermanent, but rather instruments of salvation. As later Buddhist mystics would say, *samsara is nirvana*.

One is hardly surprised to see a bow and arrow in two of Aizen's hands, so reminiscent are these attributes of the Greek godlet familiar to us, but there is also the *vajra* or thunderbolt, symbol of the theophany of sexual passion, the lotus, universal sign of tranquil ecstasy, the five-mountains staff for instantaneous travel, and the rosary ever repeating in prayer the great truths of sacrificial love. Ai-jan Ming Wang, as we would call him in Chinese Tantrism, dispels that lust which thinks of the partner as an object, calms desire pursuing its aim oblivious of the welfare of the desirable one, relieves the agony of infatuation, controls the breakup of families, and looses the bonds of selfish gratification, converting the desirous into something holy and blissful. At the same time he reveals the divine within the sexual love of human beings. When he is represented with two heads, as he sometimes is, one face is red and severe, but the other white and compassionate. In this presence one might feel at first sight a very long way away from Bemerton Parsonage and "The God of love my shepherd is," but in fact the distance is not nearly as far as superficial minds might imagine. Love is celebrated here, *karuna* or compassion for all beings, and it is love in the fullest sense, including the sexual in all its manifestations, but sanctified by the property of true love to sacrifice everything for the beloved. Therefore the truth of the Cross is here, though under forms that Western eyes have not yet learnt to understand. And so it must seem that one of the most urgent interpretations at the present day of the "love of our neighbor" of the Gospels—and indeed of the Testament of the Twelve Patriarchs too—is to recognize and embrace all peoples as participating in the great creative movement of humanity which has given us understanding and control of all things in the heavens and on and under the earth.

In another smaller hall or side-chapel on the right there is an image of Fudō Myō-ō 不動明王 (Pu-tung Ming Wang, Aryā-calanatha), another of the Five Lords of Natural Magic, with a brazier in front of it in the midst of the altar for burning quantities of incense and little logs of fragrant wood on certain days. Fudō is often shown surrounded by flames that turn above into birds, and indeed he is regarded as the protector against that tragedy which through the centuries has so often overtaken the wooden temples of China and Japan, with all their lovely carving and exquisite roofs, destruction by fire. In this presence one is moved to pray for light unchangeable and unfading, the light by which we may see our brother or sister as ourself and not as some alien being. If that light can be accepted then we may have protection from the atomic holocaust that threatens to consume all flesh, a horror hardly dreamed of so concretely by the monks who founded and built the temples on Mt. Koya. Of course love is not to be generated by fear, and men must preach the law of love for its own sake only, yet Buddhist karma, Confucian historiography, and Judeo-Christian prophecy alike have always pointed to the consequences of refusal. Seeking first the Kingdom of God today means accepting all men everywhere on a basis of absolute equality and fraternity, and seeking justice everywhere for all human needs. Only then will all other things be added— even the chance of a future for mankind, and unimaginable deepening of our understanding of the universe.

Joseph Needham

Kyoto, Aug. 1971

1

Joseph Needham and the Science of China

Derek J. de Solla Price

By the time the history of twentieth-century man has been punched into the computer, it may well be judged that to Joseph Needham lay the honor of contributing the last great traditional and comprehensive work of human scholarship. In the grand style of Toynbee's *Study of History* and Frazer's *Golden Bough,* Needham's *Science and Civilisation in China,* though little more than half complete, has become a definitive classic for our age, one of the exquisitely rare points of confluence that provide a radiant of departure for all new work.

Needham has set out to relate in seven giant volumes the whole story of the history of science and technology in China in all its social and cultural setting, from the earliest times through the coming of the Jesuits. He is engineering a transmission from one civilization to another, the like of which has never before been attempted. Not even for Western science and technology do we have such a comprehensive account, even though we live in a generation that has begun to realize that global history is dominated by the modern internationalized offshoots of that science and technology.

Science is supranational in a way that social systems and political philosophies are not, and it is perhaps for this reason that the piercing of the language barrier in this area has opened the way for a peculiarly new and powerful influence that must affect the course of subsequent scholarship in many other areas. Never before has a culture other than our own been so meticulously exposed on the dissecting board of history, and never has a teacher been so keen to make us perceive the comparisons and the contrasts in underlying structure.

Claims like these need to be made rather more forcefully than one might otherwise deem prudent, for as yet the market is hardly ready to provide a just appreciation for Needham. Our renaissance of Greco-Roman culture fell on a Europe relatively empty of literature and open to a wholesale learning of the Greek and Latin tongues; it is unlikely that our absorption of the Chinese tradition can follow such a route. Historians of science and technology still have their hands full with Galileo and Newton, medieval mechanics, and early twentieth-century quantum theory. Worse still, traditional Sinologists arrived at their competence in our educational scheme of things (as do Arabists and Byzantinists) by undergoing a process that often gives them a certificate of ignorance in things scientific. They may seize on straws of familiarity when Needham discusses the Confucians and Mohists or decides technicalities of dynastic histories, linguistics, and translations, but on the whole, the bulk of the story unfolded is still dark to them even when expressed in plain scientific terms. Even in China there are no experts who know so much of this highly technical material as well as Needham does, and even those who may know parts of the story better are laboring under difficulties. For them, too, there is a language barrier to separate them from their own history, though Needham's work has now been partially translated in China and in Taiwan.

If conditions are far from favorable for the readers, they ought to be flatly impossible for the writer. Books like this cannot normally be written, not only because of the genius demanded, but because of the unlikely combinations of genius, experience, and skill that make them possible. Who is this man who can master effectively all of science and technology, systematize the content of the vast printed classics as well as any Chinese scholar, and translate technical texts better than any Sinologist? Does it matter that he is also Master of Gonville and Caius College in Cambridge, England, a pioneer in biochemistry, an Anglo-Catholic, a very left winger in politics, the man who put the S(cience) into UNESCO, the man who testified on the use of germ warfare in China during the Korean

War, a poet of great charm and sensitivity? Of course, all of these things *do* matter, for the rarity of qualities makes the work a piece of art, as always in the finest scholarship, and the work of art depends ultimately on the personality and life of the artist.

Joseph Needham was born with the century, in 1900; his father, also Joseph, was a successful Harley Street specialist in anesthesia who had been professor of anatomy in Aberdeen before coming to London, and his mother, Alicia Montgomery, was a composer. The boy grew up in an atmosphere of intellect and culture, religiously near to the morality of the Society of Friends but also to the Oxford Movement (*not* the Oxford Group), which stressed the medieval heritage of the Church of England.It was out of this background that later, in Cambridge, he became a lay brother in an Anglican order, The Oratory of the Good Shepherd, but found eventually that the monastic life was not for him.

As these things are with English intellectuals, much of his present attitude was formed at public school. He was sent, somewhat surprisingly, to Oundle School, founded by the Grocers' Company, which catered to the industrial managerial classes and specialized, in the wisdom of its headmaster, F. W. Sanderson (a friend of H. G. Wells), in setting the boys at such things as practical laboratory work in science and in archaeology. If Wilkinson's *Manners and Customs of the Ancient Egyptians* and Schlegel's *Philosophy of History* made the first cultural impact on his mind, it was the unorthodox labs that turned him to science and sent him to Cambridge to follow in his father's footsteps as a medical student.

Needham has, however, never been able to be an orthodox anything. His brush with medicine extended as far as making him briefly a Surgeon Sub-Lieutenant in the Navy during World War I. But back in Cambridge, his tutor, Sir William Hardy, advised him against pure biology, suggesting "Atoms and Molecules, my boy! Atoms and Molecules"; and so even before his B.A. he found himself in the midst of the biochemical excitement then being led by the great Sir Frederick Gowland

Hopkins. The decade from the mid-twenties to the mid-thirties was supremely exciting, for Hoppy's laboratory, like those of Rutherford and of Bohr, was just entering a phase of sustained discovery. All Hoppy's geese, it was remarked, were swans, but they were so because Hoppy had the magic formula of personal care that transformed geese into swans by an inductive process. In 1924 Needham was elected to a Fellowship at Caius College, an institution with which his life has been interwoven ever since, up to his attainment a few years ago of the Mastership, the highest honor that his colleagues have to confer. In the same year he married Dorothy Moyle, another of Hoppy's young team, who has become an authority on the biochemistry of muscle. They are very proud of having both attained the rank of Fellow of the Royal Society (the only other married couple to have reached this distinction at the time was Victoria and Albert, and Victoria was, after all, a somewhat special case).

By the age of thirty-one Needham had published his three-volume *Chemical Embryology,* which defined a new field of study, and shortly afterward he became Sir William Dunn Reader in Biochemistry at Cambridge University. Almost immediately the historical preface of the big book took life of its own and became *A History of Embryology.* These were not, however, his first books, for he had already written much on mechanistic biology and its relations with religious and ethical thought. In England these were the years when one grew up with the biting wit of George Bernard Shaw and the leftist politics of Sidney and Beatrice Webb. J. B. S. Haldane joined the Hopkins laboratory in 1925 and must have been quite an influence politically as well as scientifically. Much influence, too, was exerted by Louis Rapkine, a Polish biochemist then resident in Paris with whom the Needhams had many seminal conversations while working together at the Roscoff Marine Biological Station.

The later thirties were years of transition and transformation for Needham. Though he continued his scientific work in bio-chemistry, he also published more and more essays on philo-sophy and Christianity and the history of science. The most important change came when three young Chinese workers entered the Hopkins lab in 1936—one of them Lu Gwei-djen, who had been sensitized to the heritage of ancient science by the work of her apothecary father. She has been one of Needham's little band of collaborators and assistants for many years.

It all went very quickly. By the end of the thirties Needham was fairly fluent in the Chinese language and making heavy inroads on classical scholarship. Shortly thereafter he set himself the big job of mastering the history of science and technology in his second land. By then the vibrations of all things Chinese were resonating violently through every fiber of Joseph Needham; he has been a complete, but never blind, Sinophile ever since, through the Kuomintang, through the war, through Mao, through everything.

As the war heightened it became clear that it was necessary to do something to strengthen the scientific and industrial liaison between China and the Allies. Needham was the obvious person to send. He duly went, and became chief of the Sino-British Science Co-operation Office from 1942 through 1946, being joined by his wife for part of the time. In a fitting personal instance of dialectical interaction, the good that Needham was able to do for China was matched by the things that this tour did for him. To give one romantic instance, after a trip by trolley to the evacuation home of Honan University in Shensi province, he writes: "I spent the afternoon exploring the library with Professor Li Hsiang-Chieh. It had been a fine collection, but successive evacuations had done some damage to the books, and the catalogue could no longer be found; there they were, many still lying tied up in bundles at the feet of the old statues of the gods, just as on the day, not so long before, when the sweating porters had dumped them down, unhooking them from their carrying-poles. Such were the circumstances in which a Cambridge biochemist was introduced by Li Hsiang-Chieh to the fact that the *Tao Tsang* (the Taoist patrology) contains a large number of alchemical works, dating from the +4th century onwards, of the greatest interest, and hardly known at all by historians of chemistry in other cultures. . . . "

It was not only intimate familiarity with the landscape and

people, language and literature, that was gained during this period. It was then that he first met the young, classically educated historian from Academia Sinica, Wang Ling, who was later able to join him in Cambridge for more than ten years of intensive collaboration.

Another outcome of the Sino-British Mission was Needham's concern with the international organization of science after the war. Through visits to Moscow and Washington and a flurry of memoranda he was successful in getting such an effort built into the emerging United Nations Cultural Organization, and when UNESCO had thus been born, he served for a couple of years as head of its Division of Natural Sciences in Paris. Then, in 1952, in the midst of intensive researching and writing before the first volume of *Science and Civilisation* went to the printer, Needham was called upon to make another visit to China, as a member of an International Scientific Commission invited by the North Korean and Chinese governments and sent by the Communist-sponsored World Peace Council to investigate the claims of the Chinese that they had been subjected to biological warfare during the Korean War. They claimed that containers of germ-carrying insects had been dropped and had led to outbreaks of plague, cholera, typhus, and anthrax. The biological evidence offered to the commission by the Chinese turned out to be quite meager; but the propaganda effects were enormous, and the personal repercussions for Needham have been determinative.

He was by far the best known and most distinguished of the scientists who composed the International Commission, and after visiting the sites, seeing what the Chinese offered as evidence, and talking with the scientists whom he already knew, he reported that there seemed to be no chicanery or deception in the charge that experiments had been made in the use of insect vectors in warfare. Needham was laughed out of court, derided for being a ready dupe of Communist propaganda, scorned for letting his known love of China lead him to the folly of subjugating science to politics. There the matter rests, for the Chinese charges were never investigated by any international body with credentials acceptable to both sides. The effects on Needham were particularly severe in view of the fact that the intellectuals in England were split down the middle over the Korean War in a way in which they are not over Vietnam. The faction made bitter enemies, and in Cambridge many faces turned against him with a lasting severity, though he still had allies among the biochemists and the support of those who really knew him at his college.

By the time he returned to Cambridge after World War II, the postwar reshuffle had disorganized his laboratory and dispersed much of it. He never really returned to biochemistry, satisfying only the statutory requirements of his Readership. Instead he became totally immersed in his big project. Huge files grew and grew. Today he sits and works in a small clearing in the rectangular disorder of his office, right at the topographic center of Cambridge, somewhat symbolically placed between the President's apple tree and the Porta Honoris of Caius College.

When the first volume of *Science and Civilisation in China* was ready to go to press, there existed already in manuscript a working text of the next three volumes and extensive notes for the rest. As it has developed, the volumes already written have grown and grown before publication so that Volume IV has now appeared in three sizable sections, each of them a volume in its own right. All along the way monographs and other separata have emerged from the prepublication research. I can never forget walking into Needham's office late in 1954 with a bright idea about a Chinese mention of medieval mechanical water clocks. It grew unbelievably, in the most hectic and intense scholarship I have ever experienced, and not too many months later Joseph, Wang Ling, and I had a viable manuscript for a full-length book on the entire tradition of *Heavenly Clockwork* (Cambridge University Press, 1959).

Now that he has caught up with himself and exhausted the material that was already (if fleetingly) complete, Needham is back in the making of Volumes V, VI, and VII. The first and last sections of Volume V have been in part held up by the difficulty of finding suitably competent collaborators in textile, mining, and ceramic technologies, but the parts which deal with alchemy are now in press. These include a survey of the background and protoscientific content of laboratory alchemy by Needham and his Chinese collaborators, and a study by Nathan Sivin (M.I.T.) of the theories which the alchemists themselves evolved to explain what they were doing. Needham also examines "internal alchemy"—or, to use his coinage, the Art of the Enchymoma—a complex of practices performed not in the laboratory but within the adept's body (similar in function but not in content to the spiritual alchemy of the West), and relates his findings to the first stirrings of biochemical thought. Volume VI, which brings Needham nearer to his starting point, on the history of the biological sciences (including agriculture and medicine), is his next priority; and this has led him into intriguing discussions, for instance, of the traditions of geobotany and soil science, where the Chinese got off to a much better start than the West. For this volume, and especially for the very difficult medical sections, Lu Gwei-djen has devoted nearly ten years of research. Such is Needham's thoroughness that Dorothy, who was always the botanist of the family on country strolls, now complains that Joseph has pre-empted her with expert talk about plant physiology and taxonomy. Volume VII now begins to recede as a summing up of the social and economic background of Chinese history wherein all comes together; it is, Needham fears, a whole additional life's work, needing much more help than he gets at present, and already, at this distance, beginning to throw off a series of independent papers and monograpis that may well bring the finished work to seven "volumes" bound actually as fifteen.

This set of volumes, produced with loving care under the direction of Peter Burbidge by the Cambridge University Press, must be one of the great monuments and landmarks in scholarly typography. The intricacies of typesetting Chinese characters, and the expense and fastidiousness of the operation, are quite incredible. Help has come most generously from the Bollingen

Foundation and from many other friends, but never, I suspect, quite enough, even though this object of humanistic support is considerably less costly than a quite tiny accelerator or computer. In the United States, Needham's operation would probably have grown into a typical academic machine, by now into its second new building, and a staff of fifty would already include a special department for the drafting of research grant applications. In England it is otherwise, and one can only wish that some means could be found to cluster just five or six scholarly full-time assistants around the great man—or three or even two!

In my estimation, the essential contribution made thus far by six volumes of *Science and Civilisation in China* lies in the systematization and presentation in English translation or summary of the substantive content of the otherwise ill-digested bulk of Chinese scientific and technical literature. Here we have the raw material on which generations of later scholarship can be founded. Here at last we have some map to tell us where to look, and some indication at least of what we shall find. In the metaphor created by André Malraux, we have been provided with a new *musée imaginaire* in which we may live constantly with a whole civilization that previously we (and that includes Chinese literary scholars, too!) knew only by special deliberate visits to the separate fragments. Even a casual browsing through any single tome is enough to break the stereotype that most Westerners will have formed about Chinese science and technology. It is not a matter of peasant-farmer and mandarin-scholar. What emerges is certainly a culture with technicalities as complex as our own, science just as deep, philosophy and experience in manipulating and transforming nature just as indissolubly bound together. Often, much the same sequence is followed in the discoveries from both traditions; sometimes one leads, sometimes the other.

There can now be no doubt that Chinese science and technology have been just as inventive, just as good, just as bad, as the science and technology of the ancient and medieval West. Now we look to the next step of wondering how it is that history does not quite act as if there is, in truth, only one natural world to discover, and that in a rather immutable order. Now that we have seen it done twice, it is striking that neither the world nor the order of discovery came out quite the same.

Then, of course, there is the perennially good talking point of why China did not have a scientific revolution and a full-scale industrial revolution like those that threw the West into dominance. Or, if you like, why was the West out of step in undergoing these curious manifestations? It is clearly a topic so admirably suited to philosophic speculation, with or without facts, that we shall never tire of giving answers and refuting them. Needham perhaps gave too much scope to the scholarly playboys by publishing Volume II on the "History of Scientific Thought" before he had let loose with his mining of the mother lode of hard stuff in later volumes dealing with mathematics, astronomy, earth sciences, physics, and mechanical, civil, and nautical engineering. I suppose it had to be done as a preamble, but I would rather stick with those content-laden texts and continue to find out more about what was actually done by the mathematicians and the engineers of that other civilization.

Among the things that the Chinese achieved must be counted, of course, the three discoveries that Francis Bacon pointed out as having been unknown to our (Western) ancients but instrumental in changing the face of the world: printing, gunpowder, and the magnet. With printing, it is after all because of this early invention in China that we can now know so much more of its antiquity than we can of the rest of the world, where the hazardous preservation of manuscript records recedes rapidly to the age that is known only from scant evidence of archaeology. As to the transmission of this crucial craft, Needham is now satisfied that Gutenberg knew of Chinese movable blockprinting, at least by hearsay.

With gunpowder, the point of excitement is that this was a Chinese invention that proceeded from pragmatic alchemical research, not from the more usual route of ingenious practical invention by artisans. By A.D. 919 gunpowder is used in flamethrowers, and by about A.D. 1000 in simple rockets, bombs, and grenades, evolving to the barrel gun by about A.D. 1120, when the Sung people were doing battle with the Chin Tartars; it is a myth of wishful thinking that the mysterious Orient used gunpowder only for fireworks! In the West the coming of this key technology of war was delayed till at least the end of the thirteenth century. Here, and again with the magnetic compass that China knew as the "South-Pointing Needle," there is clear evidence that something happens in the East, and a little later something happens in the West. The first "somethings" that happen in each case are never particularly clear or well understood, and in all cases absolutely nothing is known about any person or document actually transmitting an idea or an invention.

The putative transmissions seem, however, to cluster, almost as if they have been due to some lone intelligent traveler who came back with somewhat vague tales that were nevertheless sufficient to the cause, just as Galileo invented the telescope under stimulus from a report that a Dutchman had made a combination of lenses that enabled one to see distant things as if they were near. So it was, we must presume, with the transmission of the compass. And in that same package that arrived, maybe in the twelfth century, there came also the idea of perpetual motion machines and the germinating soul of the mechanical clock. These came out of one of the most sophisticated Chinese achievements, the long series of mechanical water clocks that included powered rotating globes and the same sort of automata beating drums and blowing trumpets that one finds today on the great astronomical clocks of the medieval European cathedrals.

Specifically Chinese are the inventions of the kite and all the practical aeronautics that go with it, and the seismograph for detecting and locating earthquakes, though with the latter there is still some uncertainty as to the actual working mechanism. Less of a mystery, but just as cunning, is the "South-Pointing Carriage," which seems to have used differential gearing to keep a figure pointing in a constant direction, no matter where the vehicle was led. But let it not be thought that one is dealing only with mechanical inventions. The other civilization had the mathematics of Pascal's triangle for binomial expansions by A.D. 1303 and Horner's method for solving equations long before Horner. It had beautifully drawn maps with accurate coordinate systems far better in detail than those of the West. In the earliest period of all, the oracle bones of the Shang period (ca. 1500 to 1000 B.C.) contain omen texts and astrology that are a rough parallel for the quite separate tradition found from the same period in the clay tablets of Mesopotamia.

Though the Chinese were highly numerate in their calendrical theory, the use of sophisticated geometrical models in planetary astronomy never became the backbone of scientific theory as it did in the West, and though there were transmissions in both directions, the two world cultures remained strikingly different in this important respect. It is, after all, the difference that led the West from Ptolemy to Kepler and Newton, and thence to Einstein. But the nonmathematical theories of chemistry and biology show the East as a fair match for the best the West could attain, and in the techniques of the industrial revolution China had scored huge successes in such things as the production of coal and cast iron, and sensible improvement of carriages and harnesses, while Europe was still inventing feudalism and monastic contemplation. One cannot, however, say these things without blowing hot and cold with pride of ownership over who did what first and who got which idea from whom. It is really a pity, for we shall have to find out much more about our past and about China's before we can make any substantial estimate of the nature and importance of the transmissions between East and West.

Presuming tentatively that our civilization survives, there must come a time when political difficulties have been sorted out and East and West coalesce into a world culture, at least to the extent that Europe long ago became an entity. It may be starry-eyed optimism to think in such terms at present, but man's quest for knowledge and understanding transcends local barriers of time and place, of language and politics. China is to

be part of our culture as we are part of its, and to Joseph Needham must be accorded, through his works, the honor of having
been able, however falteringly, to bring the two great heart-
lands of mankind together in that area where they have always
shared a common interest in the peculiar ability of man to
understand nature and bend it to his will.

2

Joseph Needham, Organic Philosopher

Shigeru Nakayama

It is not my present intention to sketch a personal profile of Joseph Needham, but rather to elucidate the characteristic way in which he interprets the history of science. In order to do so, we begin with his earlier thought as reconstructed out of his early writings and then turn to see how, in the course of his encounter with China, it provided a basis for his interpretation of the history of Chinese science. After examining various criticisms made by historians of science and sinologists as well, we shall finally try to evaluate the significance of his great work *Science and Civilisation in China.*

Needham's Earlier Thought

Joseph Needham was born in 1900. We may put it broadly that his basic ideas were formulated during the 1920s and were expressed clearly in his voluminous writings of the 1930s, on which we must depend in order to reconstruct his early outlook. Since his first visit to China during the 1940s, he has been absorbed with problems of Chinese science.

First of all, he belongs to a postwar generation—in this case the generation after the First World War, which, many historians feel, marked the greatest social transition in modern European history. In his student days, he must have been impressed by the success of the Soviet Revolution. As a scientist, he was fascinated by the publication of Engels's *Naturdialektik* in 1925. The Second International Congress of the History of Science and Technology, held at London in 1931, in which he himself was an active participant, must have been a great occasion

An earlier form of this article appeared in *Japanese Studies in the History of Science,* 1967, no. 6, pp. 138–154. This revised version is published by permission of the History of Science Society of Japan.

for him; he was strongly influenced by the thesis that "scientific theory is rooted in society," which the Russian delegation presented to that meeting. His close association with British leftist scientists of the same generation—J. B. S. Haldane, J. G. Crowther, J. D. Bernal, and others—also stimulated his concern with common problems of the day.

He began his professional scientific career as an enthusiastic follower of Frederick Gowland Hopkins.[1] In those days there existed two important scientific centers at Cambridge; one, Rutherford's Cavendish Laboratory and the other, the biochemical laboratory of Hopkins. While many ambitious young men were making careers in nuclear physics, the most exciting and best-appreciated field of contemporary science, the latter group, to which Needham belonged, steadily pioneered a solid basis for biochemistry, which was ready to flourish only after the Second World War. It is to this atmosphere that we must look for the source of his identification with the biological tradition despite the overwhelming supremacy of physical science in his youth.

It is interesting that many Marxists and sympathizers with Marxism among British scientists in the 1920s and 1930s came from the biological sciences rather than physical science. Julian Huxley and the followers of his "scientific humanism"—Lancelot Hogben, J. B. S. Haldane, and Needham—all extended their interest from biological to social problems. Some physicists, such as Bernal, sought refuge from Rutherfordian orthodox physics in a biologically and socially-minded intellectual circle. These people had a more empirical and pragmatic comprehension of Marxism than the sort of theoretical and philosophical view to which theoretical scientists (as in Japan) were often inclined.

As a biologist Needham was not, however, a stubborn defender of "neo-vitalism," on which many thinkers staked

---

[1] Joseph and Dorothy Needham, "Sir Frederick Gowland Hopkins," *British Medical Bulletin*, 1958, *5*: 299–302; Joseph Needham, "Sir Frederick Gowland Hopkins (1861–1947)," *Notes and Records of the Royal Society of London*, 1962, *17*: 117–162.

biology's *raison d'être* at the time. In a symposium of the Second International Congress of the History of Science and Technology, he, together with Hogben, stood for the "mechanist" side as opposed to J. S. Haldane's vitalism.[2] Perhaps he might have preferred to be called a materialist rather than a mechanist, as he had developed a strong antipathy toward the predominance of classical mechanistic philosphy in contemporary scientific thought.

On the other side of the coin, he had a keen affection for Whiteheadian metaphysics as outlined in the latter's *Science and the Modern World* in 1925. Although Needham was reluctant to give up the term "mechanism" entirely as Whitehead did, he must have been pleased to find in the latter's writings such statements as "Scientific progress has come to a turning point at the present time. The stable base of physical science has collapsed and, for the first time, physiology is now constructing its solid base as not a discrete but a concrete system of knowledge."[3] Needham saw in Whiteheadian organismic philosophy the synthesis of mechanism and vitalism.

Needham's personal propensity prevented him from accepting the opiate interpretation of all religious teaching, so often found in the cruder kinds of materialism. He takes a keen interest in comparative religion. Because he is undogmatic about which aspects of human activity are superstructure and which are base, he is often labeled an unorthodox Marxist. He clearly belongs to the postwar Marxist generation of the twenties, who approached Marxism through ideology and a romantic ideal, rather than through the realistic orientation toward policy and tactics that characterized the Depression thirties.

I would next propose that the unique elements in Needham's thought spring out of his viewpoint as a biological and biochemical scientist.

Let me analyze the basic constituents of his thought as

[2] Joseph Needham, "Thoughts of a Young Scientist on the Testament of an Old One," *Science and Society*, 1936, *1:* 489–495.
[3] Alfred North Whitehead, *Science and the Modern World* (Cambridge, England, 1925).

follows: (1) a strong antipathy to the mechanistic view that characterizes many physical scientists; (2) a commitment to evolution as a concept of the widest applicability; (3) Marxian dialectical materialism; (4) inclination to synthesis over analysis.

From the seventeenth century on, the mechanistic view of nature has proved its effectiveness in advancing physical science, even invading the sacred domain of biology, so that living phenomena have come to be considered reducible to physicochemical processes. As a result, biological science tended to be subordinate to physical science, but at the same time the Newtonian and later the atomistic approaches to nature did not prove readily capable of extending themselves to solve the problems of organic evolution or of historical phenomena. Biologists have made of this very field of evolution the stronghold of their independence.

Following the precedent of Herbert Spencer, Needham contemplates an evolutionary continuity of organism and human society. In his scheme, inorganic, biological, and social order are connected by an evolutionary chain.[4] From a biologist's viewpoint, organic process can be factored into successive states in time and envelopes in space.[5] Thus, just as atoms are enveloped in a molecule, "we come up from molecules to the tiniest living particles, from these to cell-constituents, from cell-constituents to cells, from cells to organs, from organs to whole animals or to the whole human body."[6] He claims as a biologist that he cannot admit a dualism of matter and mind, and that therefore no clear distinction between inorganic and organic can be recognized.

But why stop there? Needham went on to extend his evolutionary chain beyond the individual *Homo sapiens* to social

[4] Joseph Needham, "Integrative Levels: A Revaluation of the Idea of Progress," (1937), in *Time: The Refreshing River* (London, 1948), p. 235.
[5] Joseph Needham, "Biologist's View of Whitehead's Philosophy" (1941), in *Time: The Refreshing River*, pp. 180–185.
[6] Joseph Needham, "The Liquidation of Form and Matter" (1941), in *History is on Our Side* (London, 1946), p. 208.

organizations. Within the realm of social organization, family, tribe, primitive state, city state, and modern nation are arranged in an evolutionary time sequence and at the same time are successively larger, more complex, and more highly organized in envelopes of space.

To quote Needham's words, "I was unable to find any unified world-view until I took man's social life into account...."[7] "It is natural that it has fallen to biologists, psychologists and sociologists to elaborate this world-view, for they necessarily consider those levels of organization which the physicist and chemist can neglect."[8]

To each level of this evolutionary process, there is a unique approach. For instance, biological method cannot be reducible to the lower-level approaches of physiology and chemistry, nor, analogously, can sociology be based upon a biological approach alone. There is no basis for attaching judgments of value that would make the ensemble of organizational levels into a hierarchy. Only implicitly can we see in Needham's work a preference for the study of more complex and more subtle higher organisms.

Needham's thought particularly differs from that of the physical scientist in the ultimate status given to time and organization. His world is evolutionary, not the timeless Newtonian universe. Thus, his collected essays, published in 1943, were appropriately entitled *Time: The Refreshing River*. Just as life evolves from primitive to higher organisms, social organization necessarily follows a linear course of evolution. Needham finds the Marxian social stage theory as indispensable as organic evolution. He is not a social Darwinist in the Victorian heyday of British capitalism, but a man of the "postcapitalist" twentieth century. He thus cannot be expected to emphasize one aspect of evolutionary process, the "survival of the fittest," so as to enhance the merit of individualistic enterprise and free

[7] Joseph Needham, "Metamorphoses of Scepticism" (1941), in *Time: The Refreshing River*, pp. 27–28.
[8] Joseph Needham, Review of *Levels of Integration in Biological and Social Systems*, in *Science and Society*, 1943, *7:* 190.

competition. Instead, he finds "collectivism" as manifested in an ultimately worldwide Communist society to be the next higher level of organization. This is his personal version of Marxist dialectical materialism. In his most recent writing we find it much modified and deepened, particularly in the direction of independence from single-track conceptions of social development in the past, and of a critical approach to the orthodox Marxist hypothesis of a stage of Chinese history in which the fundamental relations of production were those of slave and master.[9]

He maintains a keen propensity for synthesis, which must be carried out dialectically. That dialectical materialism is the synthesis or *Aufheben* between mechanical materialism and idealism is his basic conviction.[10]

Thus religion becomes something not to be discarded in favor of materialism but to be absorbed by it. Here again, dialectical materialism is presented as the *Aufheben* of metaphysical materialism and traditional religion. "Religion begins with fear, is stabilized by priests as an instrument of subjection, and is transformed by prophets into ever higher forms of the sense of the holy." In his view, "the Communist has a more highly developed sense of the holy than any of the traditional religions."[11] Belief in the possibility of constructing an ideal society, which religion aims at, is always needed in a Communist society. Thus on a higher level, religion and Marxism must be synthesized.

In order for synthesis and *Aufheben* to be applicable, there must be some agency or mechanism to arrange and organize. One does not, of course, come to know it through structural analysis. Like Harvey before him Needham finds Aristotle more useful than the atomists. He lays stress on the role of form over that of matter, but in the postevolution-theory

[9] See his "Science and Society in East and West," *The Science of Science* (ed. Maurice Goldsmith and Alan MacKay, 1964), pp. 127–149, reprinted in Needham, *The Grand Titration* (London, 1969), pp. 190–217.
[10] Joseph Needham, "Thoughts," pp. 488–489.
[11] Needham, "Thoughts," pp. 490–491.

version of Needham, form is modified and renamed Organization; the old Aristotelian form-matter dichotomy becomes "organization-energy."[12] (Later this dichotomy is projected on Chinese thought in terms of the fundamental Neo-Confucian conceptions *li* [organization] and *ch'i* [matter-energy]. But it is organization more than energy that fascinates Needham and occupies most of his attention.) His organization is not that of the seventeenth-century momentary universe, but is temporally dynamic, expressed through process, activity, and change.

He has quoted, and has adduced the results of recent research in biochemistry and other fields to support, Whitehead's observation, "Science is taking on a new aspect which is neither purely physical nor purely biological. It is becoming the study of organisms. Biology is the study of the larger organisms, whereas physics is the study of the smaller organisms."[13]

Although it is not entirely fair to criticize Needham's views without considering the general background of thought about science in the twenties and thirties, we cannot ignore frequent specious analogies between natural and social phenomena. For instance, he quotes rather uncritically George Thomson's claim that the thought of the aristocracy tends to disintegrate and that of the masses to unite. According to Needham, while the seventeenth-century Newtonian laws of mechanics correspond to the automatic price-market, atomism is related to the selfish pursuit of profits; "atomic capitalism" has no organic connection with the whole. We are now in transition from atomistic, inorganic, and chaotic society to organized, living, planned community.[14]

His preoccupation with and faith in correspondences between the worlds of society and nature are often reflected in bold but flatly asserted theses unsupported by detailed and

---

[12] "The Liquidation of Form and Matter," in *History is on Our Side*, p. 203.
[13] Quoted from Whitehead, *Science and the Modern World*, p. 150, in Joseph Needham, "Biologist's View of Whitehead's Philosophy" (1941), in *Time: The Refreshing River*, p. 192.
[14] Joseph Needham, "Land, the Levellers and the Virtuosi" (1935), in *Time: The Refreshing River*, pp. 86–89.

critical analysis. This was a common feature of Marxist writing in the thirties.

Needham's Interpretation of the History of Science

As soon as Needham's great work *Science and Civilisation in China* began to appear, a tempest of conflicting estimates was stirred up in the world of learning. Volume I provides a more or less conventional historical and geographic background for the volumes to follow, and hence no particular occasion for debate arose out of it (although his tendency to assume that ideas or techniques that appear in two cultures have been transmitted between them rather than independently invented might be challenged if more anthropologists read him attentively). As for Volume III and the subsequent volumes, which deal with special fields of science and technology in unprecedented breadth and depth, reviewers (very few of whom are qualified both as sinologists and historians of science) are generally able only to acknowledge and admire his monumental scholarship, merely pointing out minor technical errors that are inevitable in so gigantic a work.

Most of the controversy has arisen out of Volume II, published in 1956, which deals with Chinese scientific thought. This volume strongly reflects Needham's personal interpretation of the history of science and provides evidence that his basic understanding of traditional China is in many respects a projection of the historical viewpoint developed in his youth. These two sides of his work have drawn violent criticism, the former from historians of science and the latter from sinologists. Before considering the substantive merits of this criticism, we might remark that the unfavorable criticism from some quarters—particularly in the United States, where the McCarthy purge was still in the air and Chinese studies were regarded as touchy—led to unjust doubts about the quality of the technical volumes to come. If they had preceded Volume II, it might have received less criticism.

The most critical of Needham's Marxian view of history were

the reviews of Charles C. Gillispie,[15] a historian of science, and Arthur F. Wright,[16] a sinologue.

CRITICISM BY HISTORIANS OF SCIENCE

Historians of science in the West generally cannot read Chinese and are utterly ignorant about China. Hence, when they review Needham's work, their critical eyes are focused on those aspects of the history of science in the West that Needham provides for purposes of comparison, and on his overall interpretation of the history of science.

Gillispie's argument provides a good example. His criticism may be summarized as follows: "I do not know anything about China, but I know that works on the history of Western science written by Marxist authors are not reliable, because of the nature of Marxist historiography. Needham is a Marxist, and this is a Marxist history of Chinese science, shaped from start to finish by Marxist philosophy. Therefore, Needham's interpretation of the history of Chinese science is not reliable." Gillispie thus attempts a comparison between the Marxist version of the history of Western science and Needham's analysis of Chinese science. To continue paraphrasing him, "One cannot expect respected historians of science to take a Marxist interpretation of Chinese science seriously, since its aim—to reduce the mental operations of genius to a set of economic relations—can be accomplished only through distortion. The great advances in the sophistication of the history of science in recent decades are mainly due to approaches which operate on the level of ideas—e.g., Duhem's positivism and Koyré's Platonic mathematical idealism."

If Gillispie's criticism provided any effective clue toward a new avenue of research, it would have been wholly welcomed, but his line of argument represents the thinking of most of the

[15] Charles C. Gillispie, "Perspectives," in *American Scientist*, 1957, *45*: 169–176.
[16] Arthur F. Wright, Review of Volume II of *Science and Civilisation in China*, in *American Historical Review*, 1957, *62*: 918–920.

best conventional professionals in its acuteness, its methodological finesse, and its failure to point in new directions. While it is rather easy to trace the origin of the seventeenth-century Scientific Revolution back to the scholarly writings of the Renaissance, the Middle Ages, and antiquity, artisans' contributions are difficult to evaluate, as they rarely expressed themselves clearly in writing for the public. Still, it is not wise to neglect entirely the historical significance of their tradition. In this connection, it must be appreciated that the Marxist view of history provided a stimulus to look closely at unorthodox sources. In dealing with the history of Chinese science, if we depended solely on the orthodox classification of documents according to the schemas of traditional historiography, our generalizations would have to be based on the very few fields of science that, like calendrical astronomy, were represented in the Chinese bureaucracy. The rest would remain almost entirely inaccessible. It is essential to appreciate that Needham's predilection encouraged him to enter a jungle of unorthodox and unorganized sources in such areas as technology and Taoist alchemy.

In an interesting paper, Robert Cohen contrasted Meyerson's pessimism and Needham's optimism, both formed from their studies in the history of scientific thought. This contrast is partly rooted in the fact that while Meyerson confined his exploration within the bounds of established professional science—especially physical science—to see uniform explanatory rationality, Needham used unorthodox sources which deal with the mystical apprehension of Nature, theological speculations, and craft technology.[17] Needham's optimism is based on his constant readiness to appreciate the promising possibilities of the unorthodox.

Apart from his Marxian view of history, Needham often disregards the established interpretations of traditional scholarship and provokes others with his idiosyncratic views. He is

[17] Robert S. Cohen, "Is the Philosophy of Science Germane to the History of Science? The Work of Meyerson and Needham," *Actes du dixième congrès international d'histoire des sciences* (Paris, 1964), I, 220.

not formally trained in the history of science, but he has worked out an integral and reasoned synthesis on the basis of his early intellectual commitments, never hesitating to look at old material in a new way. (For example, Chinese algebraic mathematics corresponds to organicism and Western geometrical mathematics to mechanism, Leibniz as an organicist is opposed to the atomism and mechanistic philosophy of Lucretius and Descartes, and so on.) As a result, certain of the most professional historians of science, seeing much that is unfamiliar in his discussions of the history of Western science, are unable to evaluate his work on its merits and suspect that his treatment of Chinese science must be equally distorted.[18] Perhaps it would be more useful not to ask whether his interpretations should be regarded as infallible authority but rather to make use of them as stimulating metaphors or hypotheses. Needham once quoted Otto Warburg as saying that "a hypothesis clearly expressed and capable of test [is] worth far more, even if it [is] wrong, than a quantity of vague ideas mixed up with a mass of facts in confused uncertainty." Many of his interpretations, such as the idea that Paracelsus was a socialist, are patently casual. His interpretation of the history of science on the basis of historical materialism is perhaps also to be regarded as a mere hypothesis; we will note later that it also contains contradictions.

A somewhat more just criticism is that Needham so identifies himself with his material that he overestimates the achievements of Chinese science. But this provides a healthy counterbalance in view of the ignorance of other cultures prevailing in contemporary European society. Without boundless enthusiasm, a pioneering task of such magnitude would be impossible.

One notable point of criticism common among historians of science is that Needham underestimates the role of mathe-

[18] William C. Boyd, Review of Volume II of *Science and Civilisation in China*, in *Scientific Monthly*, 1957, *84*: 213–214. One finds a greater appreciation among scholars of corresponding breadth. See, for instance, Lynn White, Review of Volume IV of *Science and Civilisation in China*, in *Isis*, 1967, *58*: 248–251.

matics.[19] Whether Marxist or not, most historians of science nowadays take it for granted that the seventeenth-century Scientific Revolution is synonymous with the establishment of a mathematico-mechanistic view of Nature. Needham, not at all enamored of mechanistic materialism, has characterized Chinese science as organismic in the Whiteheadian sense and thus, in terms of the inevitable historical trend of science, perhaps more modern than its Western counterpart. This is so shocking a statement that no one can agree with him immediately. My own view is that, without having created a unified view of, and approach to, Nature, Chinese science and technology remained an unorganized mass of fragmentary empirical knowledge lacking a nucleus up to the beginning of Westernization (or, to use Needham's word, modernization). Without a mechanistic approach, Chinese astronomy remained basically an algebraic celestial kinematics of the Babylonian type for more than two millennia. The thesis that mechanistic philosophy played an indispensable role in the establishment of modern science has become such a platitude that Needham's suggestion of other possible avenues to modern science deserves further critical examination. We can appreciate that one of his major contributions has been the application of a biologist's point of view to the history of physical sciences, yielding many fresh insights that would never occur to a historian originally trained in physics.

CRITICISM BY SINOLOGUES

The sinologist Arthur Wright doubts the validity and ultimate value of Needham's work as a whole because the arguments are based on two teleological assumptions: the development of science as a *universal progress* toward a single goal (the assumption that every ancient scientist in every country was working toward the fruition of science in the modern Western sense), and that this goal is Marxist in character (the assumption that progressive people everywhere have always aimed, whether

[19] Robert Multhauf, Review of Volume II of *Science and Civilisation in China*, in *Science*, 1956, *124*: 631.

conscious of it or not, at an ultimate collectivist utopia).

The second point of Wright's criticism is the same as Gillispie's and hence need not be considered further. Since the first point is considered by humanist scholars to be a common defect in studies of the history of science, we discuss it here.

Quoting Einstein's well-known assertion that Western science is the unique outcome of a marriage between Greek systematic formal logic and the Renaissance experimental method, Wright remarks that it is not surprising to see that science never occurred in China, where neither of these two factors developed. He thus finds fault with Needham's interpretation of Chinese culture, in which anticipations of isolated ideas or techniques of modern science are taken as the sole measure of orientation toward the future. Wright feels that this argument, which makes Taoism the forerunner of modernity, while Confucianism, Buddhism, and so on all turn out to be inhibitory, misses the point of Chinese culture. This kind of criticism is also heard from Arthur Hummel, who is generally friendly to Needham. He states that problems of the worlds of man and society rather than the natural world occupied the forefront of the Chinese mind, and that one does not gain by dismissing them out of hand. He suggests that Needham carefully evaluate the positive contributions of Chinese bureaucracy and the civil service examination system as well as their constructive influence.[20]

Unlike the younger generation of sinologists who approach China more matter-of-factly, with a better background in modern social science (and usually poorer training in Western classics and history), the old-fashioned Orientalists are usually motivated by love of the exotic and extraordinary. They find the idea of Chinese science attractive, but they generally lack the basic grasp of science, and thus hesitate to become intellectually involved, tending to either use Needham's work uncritically or reject it altogether. They feel quite at home with Chinese society, administration, and literature but are very

[20] Arthur W. Hummel, Review of Volume I of *Science and Civilisation in China*, in *American Historical Review*, 1955, *60:* 610–612.

seldom able to critically integrate science and natural philosophy into their syntheses. On the other hand, although
Needham's knowledge of and personal affection for the Chinese
people and things Chinese is considerable, he had already
developed his basic scientific viewpoint before he devoted
himself to Chinese science; his thought was coined on the
materials of Chinese science. His views on and attitudes toward
China can hardly be expected to be orthodox in the eyes of
humanistic Orientalists.

Those who work amid a humanistic tradition tend to think
of science as only a minute part of the whole culture. Even
today, despite the overwhelming claims of science, even well-
educated people seldom have an elementary grasp of its fundamentals. They are not prepared to comprehend the part it
plays in their lives. In classical Greece, in Islamic culture,
in the Middle Ages, and even in the age of the seventeenth-
century Scientific Revolution, very few people were interested
in and engaged in exact science. Until the establishment of its
legitimate position in the nineteenth century, science was only
an insignificant constituent in the ocean of general culture; its
theoretical basis was an integral part of philosophy. There was
no clear-cut scientific culture in the contemporary sense.

Neverthless, whenever enlightenment thinkers praised the
"Greek miracle" or the Renaissance revival, and when Catholic
historians more recently publicized and defended medieval
scientific achievements, their criteria for selection were based
on the viewpoint of modern science. To pluck hidden fruit out
of the bushes of several cultures and, by fitting them together,
to reconstitute the tiny strain of scientific progress; this is the
job of historians of science. If we insist on keeping all the
bushes in view, we overlook everything but the fruit that is
nearest the surface. In the case of China, while it is easy to
treat an authorized study like calendrical astronomy that was
indispensable to the Chinese political order and therefore was
amply recorded in the official histories, less obvious documents
of natural knowledge must remain in oblivion unless someone
consciously goes searching for them. Therefore, though Needham's approach has a particularly strong tone of irreversible

evolution and universal progress, his industry at picking rare fruits out of tangled bushes must be appreciated even by the most old-fashioned humanists. It is what is done with them after they have been picked that determines the historian's success at avoiding the anachronistic imposition of contemporary scientism upon antiquity and the Middle Ages.

Comments on Needham's Interpretation of Chinese Science

There are three particularly controversial aspects of Needham's interpretation of Chinese science: (1) the Taoist contribution to science; (2) Chinese organismic science as contrasted with Western mechanistic science; (3) the relation of the laws of nature and natural law. All of these are rooted in his inquiry into fostering and inhibitory factors of science and so raise basic questions not only for Chinese science but for the history of science in general. In his case, a number of insecurely based speculations, colored by his lifelong leitmotivs, have been imposed across the whole historical spectrum. They contradict not only the common views of conventional scholars, but even, unlike his more consistent earlier work, undeniable historical facts. Despite the stimulation provided by his hypotheses, several contradictions must be pointed out.

EVALUATION OF TAOISM

The following three viewpoints occupy the background of Needham's evaluation of Taoism.

1. Chinese learning is polar, the antipodes being the "masculine" Confucian thought of the ruling class and the "feminine" Taoist thought of the ruled. The elite learning of irrigation administration and calendrical astronomy is opposed to the popular techniques of chemistry, mineralogy, botany, and pharmacology; Needham, as a biologist, definitely sympathizes with the latter.

2. Taoism is antiorthodox. Science could not have flourished under Confucian feudalism but only in an opposed tradition. This is what he calls "dialectical development."

3. Science is born out of myth and magic, as in the development

of chemistry out of alchemy. Evolution is by orthogenesis. Nature-mysticism is given more weight than usual, and scholastic rationalism less.

These viewpoints are an inducement to both broaden and deepen research, making use of unorthodox sources, which hitherto have been unjustly neglected and allowed to perish. To be sure, Taoist materials do provide uncultivated but fertile land. However much it may help to develop working hypotheses, however, it would be misleading to impose uniformly upon the resultant findings the simple formula that in the development of science

$$\frac{\text{Taoist}}{\text{Confucian}} = \frac{\text{encouragement}}{\text{frustration}}.$$

In recent years, Needham has shown less of a tendency to do so. Concepts like religion, ideology, and social institutions are so essentially independent of science that it is practically impossible to construct rigorous rules about their interrelations. We may, for instance, be willing to put up with the old cliché that Confucians were interested only in moral and social affairs, and that their sober minds utterly lacked that aptitude for the abstract and fantastic which is often required for scientific accomplishment. But the fact remains that most of the great figures of Chinese science considered themselves good Confucians. Social-mindedness does not necessarily disqualify a man from being a scientist or sympathizer with science, as we can see from the examples of Leibniz, d'Alembert, Voltaire, Franklin, and Needham himself. Confucian bureaucracy, for all its tendency to frustrate the development of science, was nevertheless responsible for preserving the millennial and steadily evolving Chinese tradition of calendrical astronomy.

It would be a waste of time to attempt a net assessment simply by enumerating pros and cons. We must look more closely in order to see what aspects of science were promoted or frustrated by various systems of values. Then we must look into the whole Confucian or Taoist system to ascertain precisely which of their elements would have a given effect. Furthermore, the historical vicissitudes of the roles of these elements must be taken into consideration.

ORGANISM

Needham has an intrinsic propensity toward organism. He claims that modern science fits organism better than mechanism. "Chinese mathematical and theoretical backwardness was clothed in an organic philosophy of nature closely resembling that which modern science has been forced to adopt after three centuries of mechanical materialism." While the Westerner sees cause and effect in discrete phenomena, the Chinese perceives a nexus of situations. While time and space are treated as abstract parameters in the West, the Chinese regards them as multiplication of phenomena. Things to be grasped in terms of impact in the West are comprehended as resonance by the Chinese, and hence the latter can conceive action-at-distance more easily. However, it is very doubtful that "organism" in its Chinese version could ever take the role of a promoter of modern science. Some of the very characteristics of organism that Needham considered uniquely Chinese might be found in other premodern cultures too.

Needham seeks the elements that inhibited Chinese science. He enumerates various factors. To develop modern science, "Interest in Nature was not enough, controlled experimentation was not enough, empirical induction was not enough, eclipse-prediction and calendar-calculation were not enough —all of these the Chinese had."[21] What was missing, then, was the establishment of a mechanistic philosophy.

The present author conjectures that Needham would like to define "modern science" as a synthesis or *Aufheben* between Chinese organism and Western mechanism. According to him, modern science is neither Western nor Eastern, but universal. Mechanistic physical science as it was born in Europe could not be called modern science. It could mature into modern science

[21] Joseph Needham, "Mathematics and Science in China and the West," *Science and Society*, 1956, *20:* 343.

only once it became able to accept contributions from other cultures too.

Hence, we have to consider not only Western influence in China, but even Chinese contributions to the Scientific Revolution itself. Needham claims that the Neo-Confucian term *li* is a basic concept of natural organization, features of which were transmitted by Leibniz to the West. There they played a clearly definable part in the formation of modern science and are reflected in the world view of Whitehead. Generally, Chinese influence upon many facets of technological innovation and upon the civil service examination system have been established beyond reasonable doubt, but positive significance can hardly be imputed to Neo-Confucian influence upon the establishment of modern science. After all, Chinese *li* is a premodern type of undifferentiated "organism," while Whitehead's or Needham's is leveled organism, which emerged out of a stagnated mechanistic philosophy.

Needham's polar concept of organization/energy is applied to the Neo-Confucian dichotomy of *li* and *ch'i*. *Li* is translated as "organization" and *ch'i* as "matter-energy." Needham puts a great significance on the role of *li*, like that of the preestablished harmony of the monads in Leibniz. But it is widely recognized that the Neo-Confucian *li* was never productive in promoting scientific research even in China. Our studies in Japanese natural philosophy during the Tokugawa period show that *ch'i* played a much more productive role. *Li* is not heuristic but merely contemplative. Within the Neo-Confucian concept of *li*, moral issues and the laws of nature remained undifferentiated; thus it played an inhibitory role in the development of the modern way of thinking.

LAWS OF NATURE AND ABSOLUTISM

Needham has also presented an interesting interpretation of the relation between laws of nature and natural laws. According to him, in the Christian monotheistic tradition, a transcendental personal god is the giver of human natural law and the laws of Nature, both of which thus have a common origin. In the West under monotheism and absolute monarchy, the search for the laws of Nature has developed the same significance as obedience to natural law.

Needham's idea originates in his favorite epigram of Whitehead that modern science is an unconscious deviation from medieval theology. While Whitehead meant by this phrase only to refer to stages in the development of his own thought, Needham extended it to all of mankind as a sort of Comtean law of intellectual stages.

But what about Greek science under polytheism? Under the seventeenth-century monotheistic Spanish absolutism, we do not find a remarkable rise of science. The search for the laws of Nature was more successful in the Italian and Dutch republics than in the Middle Ages and the absolute monarchies, in the Restoration than in the Tudor period. That the Chinese were not monotheistic does not appear to have inhibited their research into Nature. It may be possible to authenticate the laws of Nature by linking them to the authority of a god or an absolute monarch, but there remain technical and metaphysical factors that cannot be reduced to ideology, and these would be more easily developed when scientists were free of the heavy hand of religious or secular authority.

Needham claims that science and democracy emerged hand in hand from a common origin in early capitalism. We may agree to the extent that the pursuit of profit in early capitalism had the same psychological basis as the pursuit of priority in scientific research. But one might just as well take as seminal the rise of individualism. As the basis of the whole complex, what matters much more than the empirical data is the world view of the historian.

Furthermore, there are some problems in accepting a parallelism between science and democracy. Democracy, originally a purely neutral term for a political institution, has attained special intrinsic value that was heightened when it had to be defended against the Fascist horror in Needham's time. During

World War II, he indefatigably preached the sterility of science under the Fascist regimes.[22] This personal experience is in itself significant. But it is difficult—except, perhaps, for an Anglo-American—to extend it to a general proposition valid for all of human history. For instance, the early nineteenth-century American masses had little enthusiasm for science, which to them smacked of European aristocracy and monarchy; the conservatism and anti-intellectualism of the majority were amply represented in the Congress. On the other hand, we have many historical instances of enlightened patrons of science. In China, perhaps more important than the issue of Taoist democracy versus Confucian bureaucracy is the problem of gerontocracy, which usually brought to nothing the natural development of individualism, and suppressed novel ideas.

Conclusion

The validity of such a question as what roles democracy and absolutism, or mysticism and rationalism, played with respect to the growth of science cannot be taken for granted, for it is essentially a macroscopic kind of questioning that often neglects to look at the inner structure of real promoting factors. It must in every case be tested; only when the parameters are positively shown to be meaningful will it become possible to correlate the macroscopic level (social structure and general trends of thought) and the microscopic level (environment of the individual researcher) and to examine their parallelism.

Criticizing Needham's view and Robert Merton's "Protestant ethic" hypotheses, Lewis S. Feuer has located the promoting factors of science in what he calls a hedonistic-libertarian atmosphere.[23] Of course, almost every scientist spends most of his time on bread-and-butter problems, and we cannot characterize any science only by the scientist's psychology.

[22] Joseph Needham, "The Nazi Attack on International Science" (1940), in *History is on Our Side*, pp. 254–298; Joseph and Dorothy Needham, eds., *Science Outpost* (London, 1948).
[23] Lewis S. Feuer, *The Scientific Intellectual* (New York and London, 1963), Chapter 8, "The Masochistic Mode of Perception in Asian Civilizations."

But Feuer's attempt, although crude and idiosyncratic, to reduce promoting or inhibitory factors to the individual level may arouse some sympathy among matter-of-fact scientists. Science is basically individual activity. The social history of science makes sense only when one's findings in that area are once more reduced to the individual level.

Borrowing Needham's terminology, the levels of the microscopic individual and the macroscopic society are entirely different in organization. For each level, a unique approach is needed. It is virtually impossible entirely to reduce one level to another, just as the national level of economic policy applies only very indirectly and incompletely to the economy of the household. The author does not believe in an evolutionary chain of necessity from the individual to organized collectivism, but in science there is an undeniable trend away from personal investigation and toward organized research at the present time, which mere correlation theory cannot explain. The ultimate problem here too is the interaction between microscopic and macroscopic levels.

Finally, for Needham's great work, D. C. Lau's comment is most adequate: "In fact it is because the book is so full of stimulating ideas that it inevitably provokes disagreement.[24] Those who do not read Chinese will have to depend on Needham's survey for decades to come, whether they like it or not. It is up to those who have access to the sources to prove or disprove his provocative ideas for themselves, and to enhance the great importance of *Science and Civilisation in China* as a determinant of our modern awareness of, and involvement with, China, by incorporating Needham's many heuristic suggestions and unelaborated ideas into strong and fruitful syntheses.

[24] D. C. Lau, Review of Volume II of *Science and Civilisation in China,* in *Nature, 178:* 1201–1202.

3

China, Europe, and the Origins of Modern Science: Needham's The Grand Titration

A. C. Graham

Why China lost its former lead and fell behind Europe is almost the first question that a layman asks about Chinese civilization, a question from which sinologists tend to shrink into their separate compartments, afraid of being caught up in inconclusive generalizations. Recently it has appeared in the context of a relatively new discipline, the history of science, presented in a more exact form that gives it a new claim on our attention. We now know that the supposed stagnation of China and the rest of Asia was illusory, their changes being slow only in relation to the accelerating development of Europe since the Renaissance, a transformation for which the only precedent is the discovery of agriculture and the transition from nomadic to settled life during the Neolithic Age. The crucial event in this process was the "Scientific Revolution" in the seventeenth century, the refining of methods of stating hypotheses in mathematical terms and testing them by controlled experiment. This was the "discovery of how to discover," the takeoff for an accelerating accumulation of knowledge, and its application to technology generated the Industrial Revolution. It seems then that we have only to ask, "Why was there a Scientific Revolution in Europe about 1600?" and "Why was there no Scientific Revolution in China or India?"—questions that look as though they are two sides of one coin.[1]

It is the second question that interests inquirers into Chinese science. I intend shortly to suggest that although the positive

This paper was published in *Asia Major*, 1971, Volume 16. Reproduced by permission of the editor and publishers.
[1] A suspicion of generalizations about nature that depend solely on authority or a priori deduction or are presented in untestable "protoscientific" forms, a recognition that the final appeal is to observation and experiment,

question is real and important there is something wrong with the negative question, but whether it is conceptually confused or not, there is no doubt that important social and cultural differences between China and the West have been brought to light by those who insist on asking it. The search for an answer has provided much of the impetus for Dr. Joseph Needham's great *Science and Civilisation in China,* and in eight papers now assembled in one volume[2] he returns to the theme again and again. In these papers, each of which displays in miniature his nearly superhuman capacity for organizing his vast store of material in the service of a lucidly argued case, the development of his thought on this problem can be followed over twenty years. *On Science and Social Change* (1944) already asks, "Why did modern science not arise in China?" and gives a fairly straightforward Marxist answer influenced by the early

---

are preconditions of modern science, but have appeared more than once in the histories of Greece, Christendom, and China without leading to a Scientific Revolution. In China the later Mohists in their writings on the sciences (ca. 300 B.C.) confine themselves to strictly testable explanations in optics, mechanics, and economics, ignore such proto-sciences as medicine, and reject the proto-scientific theory of the ascendancies of the Five Elements (Mohist Canon, B16–B31, B43). In medieval Christendom the experimental method was developed by Grosseteste (ca. 1170–1253), only to drop from sight during the fourteenth century (A. C. Crombie, *Medieval and Early Modern Science,* New York, 1959, II, 1–35, 103–211). The Scientific Revolution required not only the recognition in principle of the importance of empirical testability, but the refining of the techniques of mathematization, observation, and experiment in at least one crucial discipline. Historically the event followed the supersession of the qualitative physics of Aristotle by a quantitative and therefore strictly testable physics initially inspired in part by the sixteenth-century revival of the Pythagorean faith in number as the key to the secrets of the cosmos. See Alexandre Koyré, *Metaphysics and Measurement* (Cambridge, Mass., 1968).

[2] *The Grand Titration: Science and Society in East and West* (London, 1969). Abbreviations: *GT, The Grand Titration; SCC, Science and Civilisation in China* (7 vols. projected, Cambridge, England, 1954–). Needham has also discussed the problem of the origin of scientific method in *SCC,* III, 150–168.

I do not hesitate to apply the adjective "great" to Needham's work, although like other sinologists I am aware that his linguistic understanding is below the highest available standards. The best qualifications in both sinology and science are unlikely to meet in one person whose native language is not Chinese, and it is lucky that there is someone who has come so near to combining them.

Wittfogel: the bourgeoisie provided the setting of free and equal debate within which science can develop, but the growth of the bourgeoisie that accompanied the decay of European feudalism was not possible inside Asiatic bureaucratism. In this essay he does not yet make great claims for Chinese technology, being aware of few additions to the traditional list of Chinese inventions (gunpowder, printing, paper, the compass). His later researches revealed more and more inventions first attested in China, which he delights in listing in article after article (the mechanical clock, the driving belt, the crank, efficient equine harness, the wheelbarrow, segmental arch bridges. . . . The more recent papers recognize the technological superiority of China over most of history as a second problem; he is inclined to find the explanation in the absence of the mass chattel slavery that is commonly thought to have discouraged technological progress in Greece and Rome. By the time of *Science and Society in East and West* (1964) he gives equal weight to the questions "Why modern science had not developed in Chinese civilization (or Indian) but only in Europe" and "Why, between the first century B.C. and the fifteenth century A.D., Chinese civilization was much *more* efficient than occidental in applying human natural knowledge to practical human needs." Unsympathetic to the "internalist" approach to the history of science dominant for the last thirty years, he repeats his sociological explanation but in a much more developed and refined form, for which he acknowledges a debt to Jean Chesneaux and André Haudricourt. Although primarily interested in social and economic factors, he considers with sympathy the possibility that the genesis of modern science required the concepts of linear time and of a divine legislator, in *Time and Eastern Man* and *Human Law and the Laws of Nature.*

The researches embodied in *Science and Civilisation in China* have dispelled much of the haze which surrounded this issue. It is now clear that for most of its history the West showed no special bent toward technology. The three inventions which according to Francis Bacon had changed the face of the

world, "those three which were unknown to the ancients, and
of which the origin though recent is obscure and inglorious,
namely, printing, gunpowder and the magnet" (that is, the
magnetic compass), all reached Europe from China. Nor is it
true that the West already had science while China only had
technology. The systems based on the yin and yang and the
Five Elements which underlie Chinese alchemy, medicine, and
geomancy, do not seem to be different in kind from medieval
science, and if we prefer to speak of "proto-science" we must
apply the name to both. The greater rationality of modern

science is already present in Greek logic, geometry, and
philosophy, but for two thousand years it gave no technological
advantage for those who had it over those who had not. It is
still no doubt common to lump together Greek logic and the
modern science to which it contributed under some such
heading as "the generalized conception of scientific explanation
and of mathematical proof" or the "rational conception of the
cosmos as an orderly whole working by laws discoverable in
thought."[3] This kind of description, which rouses Needham to
polemic, illustrates, it may be suggested, the mistake of looking
for distinguishing features of a "Western civilization" conceived
as a unity nearly three thousand years old which includes Greece
and excludes Israel, instead of tracing the connections of a
"Western tradition" which is a stage, starting as far back as one
chooses to make the cut, in one of the diverging and converg-
ing lines of development that go back to Egypt and Babylon
(in which, for example, Christendom and Islam diverge out
of the late Roman civilization on which Greece and Israel
have converged). Indeed if we wish to find the best historical
perspective for looking forward toward the Scientific Revolution, there is much to be said for choosing a viewpoint not in
Greece but in the Islamic culture that from A.D. 750 reached
from Spain to Turkestan. This was the first civilization in
history that was in varying degrees the heir of all the great
civilizations of the Old World. It was in most cases the channel

[3] *GT*, pp. 42, 43.

by which Chinese inventions reached the West, but it was also the meeting place of Indian numerals, zero and algebra and Greek geometry, and of the Hellenistic and Chinese influences which ran together in the alchemy which is one of the ancestors of chemistry. A pool in which older discoveries could mix and interact, Greek, Indian, Chinese (scarcely ever Roman), was an important preliminary of the "discovery of how to discover." From about A.D. 1000 Christendom set out on the enterprise of translating the corpus of Arabic learning into Latin (including the Arabic translations of Aristotle, Euclid, Galen, Ptolemy). When the Arabic sciences passed into decline, of the three great cultures on the edges of Islam (China, India, and Christendom) it was the last that inherited its great synthesis.

The vague old question, "Why did China fall behind?" has therefore clarified and concentrated in recent decades; we might even dramatize it as "Why was Galileo born in Europe and not in China?"

First of all it is essential to define the differences between ancient and medieval science on the one hand, and modern science on the other. I make an important distinction between the two. When we say that modern science developed only in Western Europe at the time of Galileo in the late Renaissance, we mean surely that there and then alone there developed the fundamental bases of the structure of the natural sciences as we have them today, namely the application of mathematical hypotheses to Nature, the full understanding and use of the experimental method, the distinction between primary and secondary qualities, the geometrisation of space, and the acceptance of the mechanical model of reality. Hypotheses of primitive or medieval type distinguish themselves quite clearly from those of modern type. Their intrinsic and essential vagueness always made them incapable of proof or disproof, and they were prone to combine in fanciful systems of gnostic correlation. In so far as numerical figures entered into them, numbers were manipulated in forms of "numerology" or number-mysticism constructed *a priori*, not employed as the stuff of quantitative measurements compared *a posteriori*. We know the primitive and medieval Western scientific theories, the four Aristotelian elements, the four Galenical humours, the doctrines of pneumatic physiology and pathology, the sympathies and antipathies of Alexandrian protochemistry, the *tria prima* of the alchemists, and the natural philosophies of the Kabbala. We tend to know less well the corresponding theories of other civilizations, for instance the Chinese theory of the two fundamental forces Yin and Yang, or that of the five elements, or the elaborate system of symbolic correlations. In the West Leonardo da Vinci, with all his brilliant inventive genius, still inhabited this world; Galileo broke through its walls. This is why it has been said that Chinese science and technology remained until late times essentially Vincian, and that the Galilean break-through occurred only in the West. That is the first of our starting points.[4]

Among earlier contributions to the problem, Needham quotes the charming letter of Einstein to J.E. Switzer printed by Derek Price:

Development of Western Science is based on two great achievements, the invention of the formal logical system (in Euclidean geometry) by the Greek philosophers, and the discovery of the possibility to find out causal relationship by systematic experiment (Renaissance). In my opinion one has not to be astonished that the Chinese sages have not made these steps. The astonishing thing is that these discoveries were made at all.[5]

Needham takes this as a slight on Chinese civilization and springs to its defense. But Einstein does not seem to be saying anything about Chinese limitations. He seems rather to be advising Switzer not to think that a discovery was always obvious because it is now familiar, to recover the fresh eye by which it is seen to depend on a nearly miraculous conjunction of improbable circumstances. For 1400 years between Ptolemy and Copernicus the West remained satisfied with the geocentric theory although the heliocentric theory had already been proposed (and the evidence tying the motions of at least the inner planets to the sun was still available) and it forgot Hero of Alexandria's steam engine for even longer; who are we to be

[4] *GT*, pp. 14, 15.
[5] Derek J. de Solla Price, *Science since Babylon* (New Haven, 1961), p. 15, n. 10. The version quoted by Needham (*GT*, p. 43) unobtrusively smooths Einstein's English.

surprised if other civilizations failed to notice things which in retrospect seem to have been just around the corner? One does not ask why an event did *not* happen unless there was reason to expect it, and nothing in the conditions even of Europe in the sixteenth century justifies thinking of the Scientific Revolution as an event due at a certain point of maturation, as though civilization were an organism with stages which it passes through unless its development is arrested. In the absence of grounds for expectation I explain why a house did catch fire (because someone left a cigarette burning), do not go through all the other houses in turn explaining why they did not catch fire (no one was smoking, the wiring was sound, there were no bombs, no lightning). The difference follows from the fact that like effects may have unlike causes; if the event does happen we can select from the possible causes, if it does not we may not be able to enumerate all the unrealized possibilities.

If the Western development after 1600 began from a single though complex discovery, that of the means to accelerate discovery, we are concerned with an event like the invention of the wheel or of metallurgy, which we are not surprised to find diffusing from a single center where conditions for the invention are not visibly better than in many other places. We may of course find places lacking necessary conditions of an invention (the Polynesians did not invent skis because they have no snow), but for the most part it is conditions at the place of discovery that interest us. It would be pointless to ask why the Swiss did not invent skis for themselves before getting them from Norway in the nineteenth century, still more so to run over the list of maritime countries asking of each why its swimmers did not discover the crawl before its dissemination from the Pacific. But these considerations would not stop us from asking the positive question, looking for conditions favorable to the inventions in Norway and Polynesia.

The positive and negative questions are inseparable only as long as we are thinking of the difference between China and the West as one of degree. "Why is China backward?" and "Why is the West ahead?" really are two ways of putting the same vague question. We tend to suppose that this is still so when we sharpen the issue to the occurrence or nonoccurrence of the Scientific Revolution. But the questions remain two sides of the same coin only if we think of the event as having a single cause which is both necessary and sufficient, as in the more elementary kind of Marxist explanation ("Why did modern science emerge in Europe?—Because the bourgeoisie had broken free of the bonds of feudalism. What prevented it in China?—The shackling of the bourgeoisie by Asian bureaucratism"). But if Needham ever inclined to such a simplification certainly he does not now:

Whatever the individual prepossessions of Western historians of science all are necessitated to admit that from the fifteenth century A.D. onwards a complex of changes occurred; the Renaissance cannot be thought of without the Reformation, the Reformation cannot be thought of without the rise of modern science, and none of them can be thought of without the rise of capitalism, capitalist society and the decline and the disappearance of feudalism. We seem to be in the presence of a kind of organic whole, a packet of change, the analysis of which has hardly yet begun. In the end it will probably be found that all the schools, whether the Weberians or the Marxists or the believers in intellectual factors alone, will have their contributions to make.[6]

Clearly the analysis of this complex of events would not explain or need to explain why the Scientific Revolution did not occur in China. It will hardly be suggested that the spontaneous emergence of modern science in China would have required the equivalents of any of these events except, arguably, the rise of capitalism.

When we ask the negative question, we assume that there are necessary conditions for a Scientific Revolution which were present in Europe but absent in China. This assumption might conceivably turn out to be correct; but if not, how are we to enumerate all the situations in which the event could have taken place, and prove that none existed in China, and why should we wish to do so? As Einstein perceived, we are not

[6] *GT*, p. 40.

bound to ask why a civilization did *not* do something as improbable as exploring the possibilities of mathematizing its generalizations about nature and testing them by controlled experiment—a prospect less obvious, one would think, than that of a Swiss getting the idea of skis. Here it may be objected that the simple inventions we are using as analogies may be misleading us. There are scarcely any relevant preconditions, for example, of the invention of the boomerang, and scarcely any peoples on earth of whom one would wish to explain why they never got around to inventing it. But we would expect the birth of modern science to have a much more varied complex of preconditions, not the concomitant events considered in the last paragraph, but such heterogeneous factors as the meeting of Greek logic and geometry with Indian numerals and algebra, capitalism, the Judeo-Christian sense of linear time and of a cosmic legislator. However, it is precisely when factors are interrelated that it is most difficult to show that any one of them is a necessary condition. If X is ill and Y is a nurse and they meet in a London hospital it does not follow that they could not have met earlier when they were both in New York because X was well and Y was not yet a nurse. The combination of the mathematization of hypotheses and the experimental method certainly requires some mathematics and some tradition of experiment, which is enough to explain why it did not happen among Australian aboriginals; but may it not be that where these and a few other conditions are satisfied the result could follow from any number of complicated, improbable, but quite different conjunctions of circumstances?

The trouble is that explanations of China's failure to attain modern science are generally no more than proofs that she was not following the route by which we arrived at it. We are shown that one of the interlocking factors in sixteenth-century Europe was missing in China, a kind of explanation which is liable to reduce itself to the vacuous observation that conditions in sixteenth-century Europe differed from those of any other place or time. The "why not?" question could be fruitfully asked only if it should prove possible to detach the factors from their historical situation and show that they are necessary as snow is necessary to the invention of skis. We have no particular reason to expect this can be done. The problem can be seen in Needham's simplest version of his sociological argument, in the early essay *On Science and Social Change* (1944):

As we have already seen above, the rise of the merchant class to power, with their slogan of democracy, was the indispensable accompaniment and *sine qua non* of the rise of modern science in the West. But in China the scholar-gentry and their bureaucratic feudal system always effectively prevented the rise to power or seizure of the State by the merchant class, as happened elsewhere.[7]

But the rise of the merchant class would be a *sine qua non* of the rise of modern science outside Europe only if there are necessary conditions that the merchant class alone can fulfill. Are such conditions implicit in the connections that Marxists find between science and the rise of capitalism? We may instance the arguments that competing capitalists are attracted by the profitability of technical innovations, irrelevant to landowners whose income is rent; that science flourishes only in an atmosphere of free and equal debate, provided by the merchant class "with their slogan of democracy"; that the fusion of mathematics and experiment could happen only when the theoretical discoveries of Greek slaveowners were circulating among people not ashamed to work with their hands. Of course all these points are relevant to the positive question of how the Scientific Revolution came about. The close connections between science and middle-class attitudes and interests are plain enough; in English society at least science has hardly lived down its vulgar origins yet. But if we try to detach necessary conditions (a social force with a vested interest in technological advance, an atmosphere of free debate, people who could use both their minds and their hands), conditions that become vaguer the further one tries to detach them from the historical situation, it becomes less and less clear that they could be

---

[7] *GT*, p. 150.

fulfilled only by the merchant class. In any case the conditions favorable to scientific advance in a merchant class have little to do with whether or not it has won political power. Galileo after all lived at a time when the medieval fight for republican institutions in the Italian cities had long ago been lost. Asian bureaucratism did not inhibit the growth of a flourishing bourgeois culture in late imperial China: why should the political impotence of the merchant class be more of an obstacle to science than to the novel, which was already emerging in the sixteenth century? In Europe the Scientific Revolution did not wait for the seventeenth-century political struggle in England, but the novel did.

In the very interesting paper *Human Law and the Laws of Nature* (1951), Needham suggests that the concept of a divine legislator, absent in China, may have been necessary for the genesis of the idea of "laws of nature," and also for Western confidence that the secrets of a cosmos ordered by a rational being will be intelligible to rational beings. We no longer think of the phrase "laws of nature" as anything but a metaphor, but "the problem is whether the recognition of such statistical regularities and their mathematical expression could have been reached by any other road than that which Western science actually travelled."[8] Here of course we are at the crux of the matter. As with most if not all answers to the negative question we can think of alternative routes; and the trouble is not that they are plausible but that we can neither estimate their plausibility nor set limits to their proliferation. On the issue of cosmic rationality one can come to closer grips with Needham by doubting the relevance of a divine legislator to cosmic rationality even in Europe. Since Zeus gave laws only to gods and men, the Greeks should have had rather less grounds for faith in a rational universe than the Chinese, whose Heaven, however impersonally conceived, commands nature as well as man by its *ming* 命, "decree." Nothing discourages Christians from stressing the incomprehensibility

[8] *GT*, p. 330.

of a transcendent God rather than the rationality of his works, depending on how much of the Greek they have in them. But on the issue of laws of nature we are again trapped in the kind of debate in which one side suggests that there was no possibility but the one actualized and the other side produces speculative alternatives. The Neo-Confucian cosmos was rational in the sense that it reinterpreted the Heaven that one obeys and the Way that one walks as *li* 理, the pattern or layout of things, within which, wherever we discern a local arrangement, we can infer (*t'ui* 推) from one case to another. The Neo-Confucians identified *li* with the decree of Heaven and might conceivably have built a legislative metaphor on this basis, but it is difficult to see why they would need it. They were interested in laying down general principles, moral, political, and also natural, which they presented as *li;* if they had ever reached the point of formulating principles in mathematical terms and testing them by experiment, the concept of the myriad *li* which go back to one *li* would surely have provided a sufficient theoretical framework. Needham, always meticulous in collecting the relevant facts, admits that the use of the term "law" did not really catch on until after Galileo, who had spoken instead of "proportions," "ratios," "principles."[9]

This is not to deny the importance of a divine legislator in the European development. Indeed the significance of God as designer of the clockwork is clear in seventeenth- and eighteenth-century science, which inclined even after diverging from official religion to deism rather than to atheism. If there is a personal Creator, the universe is not simply there (as for Aristotle) and has not simply grown (as for the Chinese) but has been designed and constructed, so that the way to understand it is to take it to pieces like a man-made instrument and see how it works. This implies that nature is comprehensible in a special way, narrower than its rationality for the Greeks or the universality of *li* in Neo-Confucianism. Indeed the kind of rationality that seems to be guaranteed by a divine order is

[9] *GT*, p. 307. For a fuller treatment of laws of nature written in 1956, see *SCC*, II, 518–584.

partly repudiated by modern science. It denies that there are reasons for coincidences; if asked how some rare conjunction can be explained except as a warning omen, or how to account for a cruel accident without imputing injustice to God, it absolutely refuses explanation. What it requires is the treatment of a hypothesis about nature after the analogy of an instruction how to build a model, which justifies itself only when tried out, and can be tried out only if it includes exact measurements. The existence of the divine artisan authorizes the universalization of the viewpoint of the artisan, whose practice, as Needham notices elsewhere,[10] united mathematics and experiment long before 1600 but took a long time to make an impression on theory because the thinking classes do not soil their hands.

Would the absence of a Creator in Chinese thought prevent such a development? In China we find only the idea of impersonal *shen* 神, "spirit, the numinous, the divine" as the power behind the *tsao-hua* 造化, "the productive process," the process of nature by which things develop, and of a "maker of things 造物者" who is a consciously poetic personification. But it is interesting to notice how easily Chinese writers fall into this kind of language when admiring constructed models of nature such as automatic toys[11] and armillary spheres. The artificial man in a well-known story in *Lieh-tzu* 列子,[12] who seems human until taken to pieces, excites the comment: "Can man's skill then share the achievement of the author of the productive process 人之巧乃可與造化者同功乎?" The *Chin shu* 晋書, after describing how the rotation of an armillary sphere made by Chang Heng about A.D. 140 fitted the rotation of the heavens like two halves of a tally, quotes the panegyric: "His mathematics comprehended heaven and earth, his workmanship equaled the 'productive process,' his high talent and glorious art

---

[10] *SCC,* III, 158.

[11] For the kind of toy automata that presumably inspired the *Lieh-tzu* story of the artificial man, and for parallels in other cultures, see *SCC,* IV.2, 156–165.

[12] *Lieh-tzu (Ssu pu ts'ung k'an* 四部叢刊*), 5:* 7a–7b.

exactly coincided with the Divine 數術窮天地, 制作侔造化, 高才偉藝與神合契."[13] The use of such language rouses one's curiosity as to whether it occurred to anyone that man can infer how nature itself works from how his own constructions work. There is in fact a remarkable example in the comment of Chang Chan 張湛 (c. A.D. 370) on the *Lieh-tzu* passage:

近世人有言人靈因機關而生者. 何者. 造化之功至妙, 故萬品咸育, 運動無方. 人藝麤拙, 但寫載成形, 塊然而已. 至於巧極則幾乎造化.

(Recently there have been people who say that human sentience is generated through a mechanism. Why? The a-chievements of the "productive process" are extremely subtle, therefore the myriad varieties are all fostered and their activities are boundless. Man's arts are crude and clumsy, and all they can do is reproduce[14] already developed shapes in a rough way. But if human skill were perfected, it would hardly fall short of the "productive process.")

Chang Chan rejects the idea and asks: "How can it mean that a thing does not have spirit in control of it 豈謂物無神主邪?" The interest of the passage is its suggestion of a conceptual framework suitable to the development of modern science. Given an inquirer who sets out in earnest to show that something in nature works in the same way as its artificial model (such as the heart working like a pump) he would find himself drawn into measurement and experiment.

Time and Eastern Man (1964), a particularly brilliant ex-amination of the Chinese sense of time ranging from historio-graphy to clocks, is included in the volume for the sake of its discussion of the common claim that the cyclic time of Greece and India turned attention from the future while Christian eschatology encouraged a hope secularized in the doctrine of progress. Needham argues that much Chinese thought about time conceives it as linear rather than cyclic, so that the prob-lem has nothing to do with the failure to achieve modern science. The supposed links between the Scientific Revolution

[13] *Chin shu* (Pai-na 百衲 ed.), *11:* 3b7 (see pp. 115–116 of this volume).
[14] Mistranslated at this point in my *The Book of Lieh-tzǔ* (London, 1960), p. 111.

and conceptions of time are in any case so tenuous and involve so many imponderables that he offers only tentative suggestions.[15] I must confess to a personal inability to understand why the Hindu is supposed to be paralyzed by the knowledge that no human achievement can outlast a kalpa of 4,000,000,000 years, while the Christian, cramped inside a time scheme of a few thousand years from Creation to Judgment, works hopefully at sciences that have nothing to do with his salvation in the knowledge that the Last Day may already have dawned.

With regard to internal factors in the development of science, Needham shows that practical experiment without the refinement of experimental methods is common to China and medieval Europe, and that in mathematics China was strong in algebra but weak in geometry. He estimates that in the thirteenth and fourteenth centuries Chinese algebra was the most advanced in the world.[16] But the mathematics of modern science required from the beginning the entire Arabic inheritance, not only the decimal place-value system and algebra but the geometry of the Greeks. All this passed to Europe with the Arabic-Latin translations but did not reach China, although here there is a fascinating example of a historical near-miss; the Mongols brought Muslim astronomers with Arabic books, and there is evidence of a translation of Euclid in the imperial library in 1273,[17] but this knowledge never attracted attention or passed into general circulation. The mathematics developed in Europe after 1550 was an application of algebra to geometry, and Derek Price has examined an earlier nodal point in the history of science (uncovered by the researches of Neugebauer and others) at which Greek geometry had already proved itself essential.[18] The crucial discipline in the development of scientific procedures was astronomy, which even at the stage of the most primitive calendar-making com-

[15] *GT*, p. 292.
[16] *GT*, p. 44. For the Sung algebra see *SCC*, III, 38–53.
[17] *SCC*, III, 105.
[18] Price, *Science since Babylon*, pp. 1–22; Otto Neugebauer, *The Exact Sciences in Antiquity* (New York, 1969).

bines mathematization with testing (of course by observation and not by experiment); and the most important advance in mathematization before the sixteenth century was the application of geometry to astronomy by the Hellenistic school that culminated in Ptolemy (ca. A.D. 140). Behind this was the meeting of two independent traditions, Greek geometry and Babylonian astronomical observations and arithmetical computations, in Hellenized Mesopotamia after 300 B.C. (The Greeks had been weak in arithmetic as well as in astronomy.) This event, which we now see to have such decisive significance, bore no further fruit for nearly a millennium and a half, until the renewed application of geometry to astronomy by Copernicus and Kepler in the sixteenth century, followed almost at once by fusion with experimental methods. In China, as Nathan Sivin shows,[19] mathematical astronomy made a false start in arithmetical systems of simple interrelated time cycles, and interest in them soon waned as the hope was lost of reconciling them with observation. Post-Han astronomy is no longer a system, but a collection of algebraic techniques—many of them resembling the methods of Babylonian astronomy in the Hellenistic period—developed with reference only to apparent motions.[20] Needham quotes a letter in which J. D. Bernal identifies the absence of an adequate geometry to apply to astronomy as the basic weakness of Chinese science.[21] Needham is not much impressed, being more interested in external than in internal factors. We may notice, however, that this is an example which shows up particularly clearly what is involved in comparing the Chinese and Western traditions. We can examine the route by which the West arrived at the Scientific Revolution and show that China was not taking this route. But unless we wish to entangle ourselves in a demonstration that modern science could only have begun in the field of astronomy, could never have made its takeoff with laws statable in terms of traditional

[19] N. Sivin, "Cosmos and Computation in Early Chinese Mathematical Astronomy," *T'oung Pao* (Leiden), 1969, *55:* 1–73.
[20] Ibid., pp. 67, 68, 70–73.
[21] *GT*, p. 42.

Chinese mathematics, only afterwards refining its geometry to deal with astronomy, we are not even talking about the negative question which is supposed to be so important, "Why was there no Scientific Revolution in China?" The question may also be raised whether Ptolemy or even Copernicus and Kepler were in principle any nearer to modern science than the Chinese and the Maya, or indeed than the first astronomer, whoever he many have been, who allowed observation to outweigh numerological considerations of symmetry in his calculations of the month and the year. Astronomy seems to have been a mathematized discipline in which numerology was at war with observation from its very beginnings up to Kepler himself; the importance to it of geometry was merely as a model for demonstration and a tool to carry it beyond a certain point of development.

A general consideration which will occur to anyone comparing Chinese and Western thought is the much greater intellectual stringency of the latter. Granted that it is arbitrary to include Greek logic itself under the heading of scientific explanation, the importance of Greek rationality in the ancestry of modern science is not in doubt. The quality of Chinese argumentation varies with the extent of division and controversy, and it never returned to the height that it attained in the third century B.C. at the very end of the period of the competing Hundred Schools. Needham quotes the observation of H. O. H. Strange that the Greek philosopher debates with equals by logical disputation, the Chinese advises a prince with the support of historical precedents.[22] (Is it perhaps symptomatic of Needham's commitment to China that he uses the quotation not to criticize Chinese thought but to rebut the curious claim that only Europe has a sense of history?) It may be noticed that here the difference is one of degree, so that for once the positive and negative questions do come together; to the extent that the logical prowess of the West was a precondition of the Scientific Revolution, the relative weakness of China explains its failure.

[22] *GT*, p. 243.

However, people who ask why China never advanced from proto-science to science are hoping for rather more than a vague consideration which suggests that Europe would have a better chance than China. Is it possible to find some difference in kind between traditional Chinese thinking and that required by the Scientific Revolution?

The transient first impression of a fundamental strangeness, a difference in kind, does not survive a prolonged study of Chinese thought. The Chinese weigh practical advantages and disadvantages, perceive and utilize analogies, appeal to precedents, concentrate their insights in aphorisms, fascinate themselves with numerical symmetries, and sometimes reason analytically, very much as we do; if we find their thought difficult it is because of unnoticed differences in underlying concepts and in the implicit questions behind their inquiries. What we do miss, as Nathan Sivin observes, is "the notion of rigorous demonstration, of proof."[23] This concept of proof, it may be necessary to insist, is narrower than any vague idea of "Reason" that could be supposed to characterize Western thought in general. Even in the West it requires quite a special temperament to appreciate the full value of the geometrical proofs we learn as schoolchildren, demonstrations that are far in excess of the ordinary demands of common sense. Intelligent people who do not share this temperament often positively mistrust and dislike it, whether from the point of view of religious faith, romantic intuition, or Anglo-Saxon empiricism. Nor is the concern for rigorous proof equivalent to an interest in logic for its own sake. The fathers of the Scientific Revolution were interested in demonstration, not the logical forms of demonstration, and their recognition that deduction cannot lead to new knowledge compelled them to work outside the forms established as necessary by logicians. Their contempt for logic as a discipline in fact made the period from the fifteenth to the early nineteenth century a veritable Dark Age in its history, which supposed, as befits a Dark Age, that the

[23] Review of *John Fryer*, in *Technology Review*, January 1969, *71*, 3: 63.

edifice was completed by Aristotle, and forgot the advances of Stoics, Arabs, and scholastics which research is now redis-covering.[24]

It is therefore hardly profitable to make the vague accusation that Chinese thinkers lack our respect for reason, or to stress that even the later Mohists, who did study certain types of valid and invalid argument, never abstracted necessary forms like the Greek and Indian syllogism. What matters is that most Chinese thinkers (like ourselves, in most of our thinking outside the exact sciences) exchange arguments of varying and in-definite weight without seeing any point in putting premises and conclusion in the same form, filling in all steps however obvious, and pressing every line of thought toitslogical end. In particular the Chinese never developed geometrical proofs like those of Euclid, which served as the model for the demon-strations in physics of Archimedes and of Galileo and Newton. But although the ideal of rigorous demonstration has had lasting effect only in the Greek, Arabic, and Western cultures, it certainly emerged at least once in China, among the sophists and Mohists of 350–200 B.C. We may instance Mohist Canon B73, the refutation of an objection to the Mohist doctrine of universal love, which illustrates the meticulousness with which later Mohists try to put premises and conclusions in the same forms, make all logical steps explicit, and delimit what they claim to prove (in this case, merely that a position cannot be "treated as certain" or "is free from difficulty").

無窮不害兼. 說在盈否.
南者有窮則可盡, 無窮則不可盡. 有窮無窮未可智則可盡不可盡 [不可盡] 未可智, 人之盈之否未可智而 [必] 人之可盡不可盡亦未可智, 而必人之可盡愛也諄.
人若不盈 (先→) 無窮則人有窮也. 盡有窮無難. 盈無窮則無窮盡也. 盡 (有→) 無窮無難.

Canon: There being no limit is not incompatible with some-

[24] William and Martha Kneale, *The Development of Logic* (Oxford, 1962), pp. 298–378; N. Rescher, *Studies in the History of Arab Logic* (Pittsburgh, 1963).

thing being done to all. Explained by: whether it is filled or not.

Explanation: (Objection) If the South has a limit it is exhaustible, if it has no limit it is inexhaustible. If whether it is limited or not is not yet knowable, then whether it is exhaustible or not is not yet knowable, whether men fill it or not is not yet knowable, whether men are exhaustible or not is likewise not yet knowable, and it is erroneous to treat it as certain that men can be exhaustively loved.

(Answer) If men do not fill the limitless then men are limited, and there is no difficulty about exhausting the limited. If they do fill the limitless then the limitless has been exhausted, and there is no difficulty about exhausting the limitless.

As for the Chinese language, Needham is content to expose the fallacy that the script, mistakenly supposed to be not logographic but ideographic, would inhibit abstract thought, and to point out that the exposition of twentieth-century science in Chinese has presented only the problem common to all languages of evolving a technical terminology.[25] I have myself argued elsewhere that claims that Chinese thought is hampered by confusions over distinctions marked by Indo-European number and case always break down when a concrete instance is offered, but also that the discovery of logic as an independent discipline (a dispensable luxury for the Scientific Revolution, as we have seen) may be easier in an inflected than in an isolating language.[26] Logically the advantage of an inflected language is that the changing word forms illuminate the organization of the sentence, an advantage which has nothing to do with the supposed utility of the distinctions marked, which may be quite irrational (as with gender). The structure of an isolating language is invisible without the aid of modern linguistics and offers no foothold for an exploration of grammar or logic; it allows any degree of exactness or inexactness, so that the vagueness or precision of Chinese thinking must always be

[25] *GT*, pp. 37–39.
[26] "The Logic of the Mohist Hsiao-ch'ü," *T'oung Pao,* 1964, vol. 51 (Part 4, "The Mohist Logic and the Chinese Language"), pp. 39–53.

attributed to extralinguistic factors. The sharpening philosophical controversies of the fourth and third centuries B.C. involved a clarification of terminology and tightening of syntax in some ways comparable with the effects of science on contemporary Chinese. For example, because of problems raised by controversy over their doctrine of universal love the later Mohists were interested in quantification of the object, and they defined two of the quantifiers (盡, 莫不然也, " 'All' is 'none not' "; 或, 不盡也, " 'Some' is 'not all' "). In the service of this concern they regularized the grammar of the distributives. To refer back to the subject they use *chü* 俱, "all," *huo* 或, "some," *mo* 莫, "none"; to refer forward to the object, *chin* 盡 and the patterns 有, 無 . . . 於 . . . (有愛於人, "love some men" 無遺於其害也, "overlook none of the harm in it"). In other constructions for which the last formula could be mistaken, confusion is avoided by using *huo-che* 或者, *wu* 毋, and the preposition *hu* 乎; 或者遺乎其問也, "Some are overlooked by his question"; 心毋空乎內, "the heart has no empty space inside it." In the language of late Mohist dialectics, vocabulary is regularized (a fact obscured by great graphic variety due to imperfect adaptation of graphs to later usage), there are virtually no synonyms among particles, idiom is avoided, syntactic consistency is observed even at the cost of sentences so extraordinary that they have generally been taken to be corrupt (有有於秦馬, "have some Ch'in horses").[27] Given the extralinguistic conditions for the development of modern science, the Chinese language would presumably have adapted itself much as seventeenth-century English allowed itself to be reformed by the Royal Society.

An important point of Needham's, further developed in *Science and Civilisation in China*,[28] is that medieval science or proto-science with its Galenic humors in Europe and yin and yang and Five Elements in China is culture-bound, but from the point that science is mathematized and experimentally

[27] For the references see Graham, "The Logic of the Mohist Hsiao-ch'ü," pp. 6, 11–14.
[28] *GT*, pp. 15, 16; *SCC*, III, 447–451.

testable it acquires the cultural universality of mathematics and logic. There is nothing in our culture that carries so openly the marks of its Oriental origin as the numerals that we still call "Arabic" in contrast with "Roman," or the concepts which still bear Arabic names—algebra, zero, zenith, nadir, chemistry—but since they belong to culture-free disciplines we do not feel them to be alien at all. There is no reason to assume that the world will keep for long its feeling that modern science is specifically Western. The geographical region where modern science began remains important only as long as it keeps the initial advantage of having been the discoverer, but afterwards presumably will concern historians alone, like the origin of agriculture in the Middle East and, within the already industrialized world, the origin of industrialism in England. To think of the modernization of Asia and Africa as their "Westernization" in any but a short-term sense is to forget that the Industrial Revolution disrupts and transforms all preceding cultures in West and East alike, and at the same time throws their resources into a common pool. It is possible to wonder whether we ourselves will necessarily be classed as belonging to a "Western civilization" by historians of the not so far future. They may find it more convenient to treat the West as the first of the great agrarian civilizations to lose its identity after the Industrial Revolution. If we knew more about the tribes that first settled on the banks of the Nile we might find cultural continuities comparable to those between medieval Europe and ourselves, but we should not be tempted to regard the revolutionary change to agriculture as a mere episode in an Egyptian tradition. Such assumptions of a surviving homogeneous culture as that a Westerner, whatever his overt beliefs, has a sensibility rooted in Christian symbolism that allows only a superficial conversion to Vedantism or Buddhism, and a coherent artistic heritage from the Renaissance that admits Oriental influences only at the level of the picturesque, no longer seem self-evident as they did even a generation ago. The whole European and Middle Eastern conception of religion as the pursuit of moral improvement in the service of a personal and transcendent God seems to come less and less naturally to the spiritually hungry even when they are professing Christians, which suggests as profound a break in a cultural succession as it would be possible to conceive. However we may judge the alien contributions during the last century and a half to every aspect of our culture outside the immediate reach of science, from Schopenhauer's debt to the Upanishads to the Black American and now African and Indian elements in popular music, it is already obvious that more is involved than the mere exoticism of eighteenth-century chinoiserie. The Japanese woodblock for the Impressionists and Japanese architecture for Frank Lloyd Wright, Chinese poetry for the Imagists and African sculpture for Picasso, were active influences at crucial moments in the development of major modern styles.

It is not altogether easy to break the habit of thinking of history as blindly groping toward a goal that the West alone was clever enough to reach, and Needham himself sometimes has the air of making allowances for the Chinese and offering compensations. But the only conscious goal that anyone has been able to find in the social processes that led to modern science is capitalist profit. Accidents such as Greek geometry encountering the Babylonian astronomy which it was to transform or China (which had the astronomy) developing algebra instead of geometry, are hardly to the credit or discredit of a civilization. When we consider how slowly both the West and China have responded to alien discoveries as long as they were confident of their own cultural superiority (the Indian numerals adduced by the Syrian bishop Sebokht in 662 as proof that the Greeks do not know everything, but their use in Europe unattested until 976, after which they only very slowly superseded Roman numerals; Euclid unnoticed in China until the Jesuit translation of 1607 although apparently available from Muslim astronomers as early as 1273),[29] we can see a direct connection between the superiority of the West

[29] *SCC,* I, 220; III, 52, 105, 146.

about 1600 and its abject inferiority about 1000, which forced it to borrow the Arabic sciences wholesale and thus become the possessor of the all-important combination of Greek and Indian mathematics. Is it necessary to say more than that one set of conditions for the genesis of modern science came together in sixteenth-century Europe, and that since it spread too fast to allow independent occurrence elsewhere this is the only set of conditions of which we can ever know? The tremendous dynamic of the Scientific Revolution distinguishes it in this respect from the only comparable episode in history, the Neolithic invention of agriculture and the ensuing urban revolution in the Middle East. Agriculture continued to spread through the millennia between the natural barriers of the Atlantic and Pacific, so that there could be time and space for its independent discovery elsewhere. But the few centuries of the spread of modern science, although long in terms of its own accelerated time scale, are short in relation to the slow rise and fall of agrarian civilizations.

Nathan Sivin begins his recent book on Chinese alchemy with the observation that to ask of Chinese science, "Why did it not spontaneously evolve into modern science?" is a question best postponed until more is known. But he does not doubt its importance:

This question, to be sure, is crucially important, for much of China's convulsive experience of the past century or so, and indeed much of her predictably convulsive experience of decades to come, are part of a world upheaval in which intellectual, social and economic consequences of the Scientific Revolution are gradually asserting themselves.[30]

But here we are concerned with something different, the factors in Chinese society and culture favorable or unfavorable to the assimilation of industrial civilization in all its aspects, and the problem of origins is left behind. If we imagine sixteenth-century Europe invaded from outside by electronics and plastics, air travel, nuclear energy and napalm, television and pop

[30] Sivin, *Chinese Alchemy: Preliminary Studies* (Cambridge, Mass., 1968), pp. 1, 2.

music, its struggle to adapt would not be eased by being itself on the verge of the discovery of quantitative physical science. Whatever China's problems, absence of conditions in which mathematization could combine with the experimental method is no longer one of them. That tradition of centralized bureaucracy which according to Wittfogel and Needham inhibited the growth of the merchant class and therefore of science, may have turned to China's advantage, since as soon as science is visibly a means to power a state's fear of more modern states becomes a stronger motive for importing it than commercial profit. The un-Chinese concept of a divine legislator or watchmaker has long ago lost its utility. Needham himself has often emphasized that a tendency toward organic rather than mechanistic thinking, although it conflicted with the presuppositions with which modern science began, may facilitate the assimilation of twentieth-century science. Here one is again reminded of the difficulty of throwing off the assumptions of the old vague question, "Why did China fall behind?," even when the issue has narrowed to the presence or absence of certain conditions immediately preceding Galileo. If the historians of science are right in so concentrating the issue, the setting of the Scientific Revolution in Europe becomes a matter of particular conditions, some persistent (such as the habit of philosophical and theological logic-chopping) and others transient, and we can no longer assume that outside them Western civilization in 1600 was any less remote than China from a civilization already revolutionized by modern science. It is irrelevant that the conflict between traditional culture and the Scientific Revolution has been so much weaker in the West than elsewhere. The civilization that first advances from proto-science to science will have only the problem of adapting to the Scientific Revolution itself; all others must adapt also to the alien civilization from which it reaches them, which is less and less like themselves or any other agrarian civilization, including the Europe of the past.

4

The Chinese Concept of Nature

Mitukuni Yosida

1. Law and Change

Toward the end of the Astrological Treatise ("T'ien kuan shu 天官書") of the *Shih chi* 史記, the first of China's long tradition of Standard Histories (ca. 90 B.C.), the astronomer-historian Ssu-ma Ch'ien 司馬遷 briefly expresses his view of the relationship of man and nature, one of the perennial themes of early Chinese thought:

Since the beginning, when humankind came into being, rulers in successive eras have observed the motions of the sun, moon and stars. Through the reigns of the Five Emperors and Three Kings [that is, throughout antiquity], as the effort was continued their knowledge became clearer. [China, the land of] the ceremonial cap and belt, was considered "inside," and [the land of] the other peoples considered "outside." The Middle Lands were divided into twelve provinces. Looking up, they contemplated the signs in the sky. Looking downward, they found analogues to these on the earth. In the sky there were the sun and moon, on earth yin and yang. In the sky there were the Five Planets, on earth the Five Elements. In the sky there were the lunar mansions, on earth the territorial divisions. The Three Luminaries [that is, sun, moon, and planets] are the seminal *ch'i* of yin and yang. The *ch'i* originally resides on earth, and the Sages unify and organize it. Since the time of Kings Yu and Li 幽厲 of the Chou era, the ruling house of each state used different means of divination to find a way to conform to the exigencies of the time as manifested in celestial omens. From their documents and books no rules whatever for portents can be extracted. Therefore when Confucius laid out the Six Classics, he merely recorded abnormal phenomena, and did not write down interpretations. As for the Way of Heaven and the destiny of man, he did not transmit [his thoughts]. The person capable of receiving [this knowledge] does not need to have it spelled out for him; even if it is spelled out, an unfit person will not understand it.[1]

Translated by Henry Mittwer, S. Nakayama, and N. Sivin.
[1] *Shih chi*, ch. 27.

The ideas of an analogy between earth and sky, and of omens as a link between nature and man, here emerge from an archaic background in which Heaven is seen as personal and all-powerful.

It was Heaven, the so-called supreme spirit on high, which brought forth people and made nature for their provision. That faculty is already reflected clearly in the oracle-bone documents of the Shang era, in the mid-second millennium B.C.:

Divined on [a certain] day, [unidentifiable] divining. Will the Lord command much rainfall in this third month?

Divined on the forty-seventh day of the sexagenary cycle. Will the Lord send down drought?

Divined. Will the Lord command rain to make the harvest come at the correct time?[2]

As is well known, this oracle-bone writing was the script used in divinations addressed to the celestial Lord and spirits. This Lord held the power to make the rain fall, bring drought, and control the abundance of the year's harvest. He was the powerful spirit who controls nature.

At the same time he influences human conduct:

This spring when the King makes a tour of his hands, will the Lord give us his protection?[3]

Will the Lord sanction the King's establishment of a city?[4]

In this manner the diviner makes his inquiries of Heaven. When the King mounts an expedition against another state, or establishes a capital, he must always obtain the approval of the celestial spirits. This being so, from time to time the Lord will visit calamity upon the human world.

Will the Lord send down misfortune?[5]

---

[2] Lo Chen-yü 羅振玉, *Yin hsu shu ch'i ch'ien pien* 殷虛書契前編 (Documents from the Yin wastes, first collection; Peking, 1912), 3.18.5, 3.24.4, and 1.50.1.

[3] Lo, *Yin hsu shu ch'i hsu pien* 續編 (Documents from the Yin wastes, sequel; Peking, 1933), 5.14.14.

[4] Lo, *Yin hsu shu ch'i hou pien* 後編 (Documents from the Yin wastes, second collection; Peking, 1916), *hsia*, 16.17.

[5] Shang Ch'eng-tso 商承祚, *Yin ch'i i-ts'un* 殷契佚存 (Extant Yin records; Peking, 1933), p. 36.

Truly the celestial Lord and spirits possess strong power to control all things.

Every sort of natural phenomenon manifested by this mighty Heaven is in itself an expression of Heaven's will. Thus the ruler must depend upon observing the signs in the sky in order to infer the will of Heaven with respect to the world which he rules. And insofar as mundane human conduct was controlled by Heaven, necessarily through human conduct as well the will of Heaven is known. Thus nature and human society, inseparable, coexist in accord with the laws by which Heaven exerts its control. These laws manifest themselves in the sky as the regular rotations of the sun, moon, and planets, and on earth as the principles of yin and yang and the Five Elements. The Astrological Treatise of the *Shih chi* is a kind of book of divination. From sun, moon, planets, and fixed stars to such meteorological phenomena as clouds and winds: all were celestial revelations for the prediction or assessment of human conduct.

Now the terrestrial yin-yang principles are principles of transformation. They are explained by the Changes—the system of thought of the Book of Changes—which evolved out of ancient divination:

> What cannot be measured by yin and yang is called spirit. The Changes is equivalent to heaven and earth.[6]

Complete foreknowledge of transformation is not a prerogative of human society. To know the impossibility of this foreknowledge is to know the spirit. Thus the Changes is identical with heaven and earth, that is to say, with nature. Nature is transformation, and this transformation arises out of the two *ch'i* of yin and yang.

> The activity of the spirit shows itself in the totality of phenomena; this is transformation.[7]

The situation in which a balance of the two *ch'i* is maintained is the ideal for human beings.

---

[6] *I ching* 易經, "Hsi tz'u 繫辭," 1.5 and 1.3 respectively. Cf. R. Wilhelm, *The I Ching or Book of Changes* (tr. C. F. Baynes; Bollingen Series XIX: New York, 1950), pp. 323 and 315.

[7] *I ching*, "Shuo kua 說卦," 5. Cf. Wilhelm, p. 287.

One yin and one yang [that is, their alternation] is called the Way. What makes the process continue to completion is the original moral nature of mankind.[8]

In this way the principle of natural change has been extended to apply to human society. There too the balance of yin and yang was considered the moral ideal. The principles of transformation can be applied as they are directly to the human world, bringing it into correspondence with the Way of Heaven, but these principles can be known only by transcendent knowledge.

Search out the beginning, return to the end; thus you will know life and death.[9]

Human life and death were also included in the natural principles of beginning and end.

Thus the terrestrial aspect of the laws of transformation is manifested as the Five Elements:

Men follow the laws revealed in the celestial signs, living in accord with the nature of terrestrial things. Heaven and earth give rise to the Six *Ch'i* [yin and yang, wind and rain, dark and light], and from these are born the Five Elements [Metal, Wood, Water, Fire, and Earth]. Out of man's use of these come the Five Flavors [sour, salty, acrid, bitter, sweet], the Five Colors [virid, yellow, scarlet, white, black], and the Five Modes [in music]. But when these are indulged to excess, confusion arises and in the end man loses sight of his original nature.[10]

The Five Elements, extended to become the Five Colors, Five Modes, and other fivefold categorizations, took a central role in human life; this is affirmed by their importance in ceremony.

"Heaven has given rise to the Five Materials, and all the people use them. Not one of them can be done away with."[11] The Five Materials, which correspond to the Five Elements, are five substances necessary for human life.

The substances that man employs were originally produced by Heaven, the supreme god. Onto this idea was projected the

8 *I ching*, "Hsi tz'u," 1.5. Cf. Wilhelm, p. 319.
9 *Ibid.*, 1.3. Cf. Wilhelm, p. 316.
10 *Tso Chuan* 左傳, Duke Chao, year 25.
11 *Tso Chuan*, Duke Hsiang, year 27.

concept of the earth, on which the various things exist. The result was extended to the notion that by the functioning of *ch'i,* which belongs to the phenomenal world, everything within the realm of human experience, from the Five Elements to the Five Modes, came into being. The conception of the Five Materials is from another viewpoint the idea that everything that participates in the making of this world was formed out of five basic substances. The Five Materials are the five basic constituents of things, and thus the fundamental elements. These elements must be intimately associated with all phenomena and influence them. Many sets of things were classified according to this influence or correspondence. In principle all things could be so distributed. At the same time, Heaven remained the intermediary for the relationship of Nature and man.

Some examples of Five-Elements classification are shown in Table 4.1, on the following page.

This form of classification was practically complete by the beginning of the Ch'in era, as can be seen in the Monthly Ordinances ("Yueh ling 月令") chapter of the *Li chi* 禮記 (Record of rites), and was further enlarged in the Han period. The Five Elements were not, however, considered fixed and static; nature is vicissitudinary, and its changes are reflected in certain recurring relationships of the fundamental elements. One such relationship was the Mutual Production (*hsiang sheng* 相生) succession. This system followed the order of elements which appears in the Monthly Ordinances, linking them as phases of a creative cyclical process by which Wood gives rise to Fire, Fire yields Earth in the form of ash, and Earth produces Metal as ores grow inside the terrestrial womb. Another important postulate was the Mutual Conquest order, wherein Fire conquers Metal by melting it, Metal overcomes Wood by cutting and carving it, and Wood defeats Earth whether by digging it up or growing out of it. A great variety of different dynamic relationships, from physical change to human affairs, can be accounted for by choosing either the Mutual Production or Mutual Conquest order.

Table 4.1. Five-Elements Categories

Category	Five Elements				
	Wood 木	Fire 火	Earth 土	Metal 金	Water 水
Calendar Signs	*Chia* 1 *I* 2	*Ping* 3 *Ting* 4	*Wu* 5 *Chi* 6	*Keng* 7 *Hsin* 8	*Jen* 9 *Kuei* 10
Seasons	spring	summer	between summer and autumn	autumn	winter
Directions	east	south	center	west	north
Emperors	Fu-hsi	Shen-nung	Huang Ti	Shao-hao	Chuan-hsu
Musical Modes	*Chueh*	*Chih*	*Kung*	*Shang*	*Yü*
Tastes	sour	bitter	sweet	salt	acrid
Internal Organs (Alternate Orders)	liver spleen	heart lung	spleen heart	lung liver	kidney kidney
Colors	virid	red	yellow	white	black
Human Faculties	demeanor	speech	vision	hearing	thought
Virtues	charity, benevolence	courtesy, propriety	wisdom	justice	fidelity, sincerity
Creatures	feathered creatures	hairy creatures	fleshed creatures (humans)	shelled creatures	scaly creatures

This way of thinking, which originated with Tsou Yen 騶衍, became stronger after the Han dynasty. Every aspect of nature and society was classified and interpreted accordingly. Since the succession orders were cyclical, this meant a widened emphasis on recurrent dynamic relations in Chinese thought. Nature was change, and that change remained controlled by Heaven, now envisioned as an objective and organic order.

2. Ch'i and Nature

The realm of nature changes constantly, but its coherence

comes from the principle of governance by the celestial order. But how did that universal order come into existence?

The idea of a pneumatic *ch'i* 氣 as the origin of the universe was expressed in the Disquisition on Astrology ("T'ien wen hsun 天文訓") of the *Huai nan tzu* 淮南子 (ca. 120 B.C.). In the beginning, nothing had physical shape, and the first spontaneous formations were the continua of space and time (*yü chou* 宇宙). Out of these were produced the original *ch'i*. This *ch'i* was heavy and stable, but its lighter part rose and became the sky. The heavy and turbid part gathered and became the earth. The gathering of the heavy substance took time, and hence the sky was formed earlier. Then the *ch'i* of sky and earth met and became yin and yang. The active *ch'i* of yin and yang became the four seasons, and as this *ch'i* of the seasons scattered it formed the various phenomenal things of the earth. The hot *ch'i* of yang gathered and became Fire. Next, the essence of the *ch'i* of Fire became the sun. The cold *ch'i* of yin gathered and became Water. The essence of the *ch'i* of Water became the moon. The encounter of the *ch'i* of the sun and moon gave rise to the stars. This is the celebrated cosmogony of the *Huai nan tzu*.

The idea of the beginning of the universe in an undifferentiated Chaos was set forth early by Lao-tzu 老子. He equated this chaos with the Tao. "Shape without Shape, Form without Objects"[12] is the Tao, the origin and principle of all things:

There is a thing confusedly formed,
Born before heaven and earth.
Silent and void
It stands alone and does not change,
Goes round and does not weary,
It is capable of being the mother of the world.
I know not its name
So I style it "the way."
. . . .
Man models himself on earth,
Earth on heaven,

[12] *Lao tzu*, 14.

Heaven on the way,
And the way on that which is naturally so.[13]

This is not a creation, either ex nihilo or by a creative "chemical" transformation of one substance into other distinct substances, but rather a continuous physical process described by the artisanal term "fabrication" (*tsao wu* 造物).

Next Lao-tzu says "Tao produced the One, the One produced the Two, the Two produced the Three, and the Three produced the myriad phenomenal things."[14] Thus the Tao is one of the basic manifestations of the *ch'i*, and from it the yin-yang duality arises. Yin and yang combine and become "the three," or plurality, and from plurality arises individual things. The phenomenal world is continuously produced out of the unitary state of *ch'i* called the Tao. This is not a purely sequential process in which entirely different orders are created out of the Chaos and developed step by step. The Chaos, without changing its nature, takes on form as sky and earth, and then as yin and yang.

It was the *Lieh-tzu* 烈子 (probably compiled ca. A.D. 300) which carried these ideas further, saying that "the Unborn can give birth to the born, the Unchanging can change the changing."[15] The original state from which the universe was formed is prior to all determination, and the words "creation" and "extinction" are inapplicable to it. In regard to the evolution of all things out of the ineffable, this book made use of four phases of definition:

There was a Primal Simplicity, there was a Primal Commencement, there were Primal Beginnings, there was a Primal Material. The Primal Simplicity preceded the appearance of the *ch'i*. The Primal Commencement was the inception of the *ch'i*. The Primal Beginnings were the *ch'i* beginning to assume shape. The Primal Material was the *ch'i* when it had begun to assume substance. *Ch'i*, shape, and substance were complete,

---

[13] *Lao tzu*, 25, tr. D. C. Lau, *Lao tzu. Tao te ching* (Harmondsworth, Middlesex, 1963), p. 82.
[14] *Lao tzu* 42.
[15] *Lieh tzu*, ch. 1, tr. A. C. Graham. *The Book of Lieh-tzŭ* (London, 1960), p. 17.

but things were not yet separated from each other, hence the name Chaos. Chaos means that the myriad things were confounded and not yet separated from each other.[16]

At this point Heaven and Earth spontaneously separate. Man comes into being, and all other phenomenal beings and things with him. But this life and change have as their substratum nonbecoming and nonchange. The most basic *ch'i* of the universe is everlasting and unchanging. There is neither increase nor decrease, and no alteration.

Turning without end
Heaven and Earth shift secretly.
Who is aware of it?
So the thing which is deficient in one place is replete in another; the thing which is waxing here is waning there. Deficient and replete, waxing and waning, it is being born at the same time that it is dying. The interval between the coming and the going is imperceptible; who is aware of it?[17]

The recognition that man's key role in the universal order made him superior to other phenomenal beings was due to Tung Chung-shu 董仲舒 (179–104 B.C.). As he put it, "the *ch'i* of Heaven is above, the *ch'i* of earth is below, and the *ch'i* of man is in between."[18] Human beings are not only located between Heaven and Earth, but are in correspondence with the natural order. To exemplify this principle, Tung likened the 360 joints in the human body with the 360 degrees by which celestial positions were marked. The flesh and bones of the human body corresponded to the texture of the earth. The ears and eyes corresponded to the Sun and Moon, and the five internal organs to the Five Elements. The four limbs were the Four Seasons. Man was a microcosm in which all the contents of the macrocosm were reflected. Tung also broadened the range of Five-Elements thought by applying the dynamics of the succession orders to every aspect of experience, emphatically including statecraft. Administrative posts, official buildings,

---

[16] *Lieh tzu*, ch. 1, tr. Graham, pp. 18–19, modified.

[17] *Lieh tzu*, ch. 1, tr. Graham, p. 27, modified.

[18] *Ch'un-ch'iu fan lu* 春秋繁露 (Copious dew on the spring and autumn annals), 56.

and palaces in the Han period were named for stars and asterisms in order to ensure the correspondence of the political order with that of Nature. Integrated was the idea that by observing anomalies in the sky one could detect the corresponding deficiencies in human society. Under Tung's urging the long-established official sponsorship of philosophy was restricted to Confucianism. Although its new emphasis on cosmology would hardly have been approved by Confucius, this Han Confucian synthesis did undercut the intellectual appeal of other philosophies that vied for sponsorship.

The thought of Confucius himself was humanistic, in the sense that he felt the problem of the good life was ethical and could be solved only by the improvement of human society. He talked about Heaven and the Celestial Way, but his was not the sort of Heaven that could serve as the original principle of the universe, like that of Lao-tzu and Lieh-tzu. The universe was practically perceived only through its influence upon man; its transcendental nature was not denied but was beside the point.

At a time when the bureaucratic state was taking shape under the influence of Confucianism, Wang Ch'ung 王充 (A.D. 27–97), the author of the *Lun heng* 論衡 (Discourses weighed in the balance), took issue with some of its basic assumptions. Wang did not propound a philosophy of his own, but skeptically challenged the beliefs of his time on the basis of a combination of observation and hearsay. According to him, Heaven and Earth have no will. Man's appearance on this world was due to a mere fortuitous concourse of the *ch'i* of Heaven and Earth.

The meeting of the *ch'i* of a human couple that produces offspring comes about as a result of the workings of sexual desire rather than because of the couple's wish for progeny. Babies are born by a spontaneous process of which the parents are only the agents. "Heaven is idle and spontaneous." Thus the idea that natural calamities are warnings from Heaven must be wrong.

A sickness of the body is like a calamity from Heaven. When the blood and pulse are not in equitable adjustment, man contracts a disease, and when the winds and the *ch'i* are not in harmony, the year develops calamities. If calamities are Heaven's reprimands to the administration of the state, are diseases also Heaven's reprimands to individuals?[19]

Wang Ch'ung also rejected the sorts of physical thought that his predecessors and contemporaries (Tung Chung-shu especially) were building on the foundation of the Five Elements and their correspondences and succession orders. Wang found the idea of a correlation between the natural and social orders completely unacceptable. He also argued against prevalent notions of soul and spirit. His monistic Primal *Ch'i* (*yuan ch'i* 元氣) was at the origin of all things, and psychic as well as material entities became mere aspects or embodiments of it. Man was locked into a fixed destiny, which he could not alter by his striving. Since the Way of Heaven is utterly spontaneous and devoid of will, physical change is in principle entirely susceptible of materialistic and mechanistic explanation.

It is well known that a new era of Chinese thought flowered in the Sung era. Post-Han Confucian theory had largely been confined to questions of morality, with little attention to natural philosophy. The Taoist and Buddhist schools had remained on the level of high metaphysics. But in the Sung these currents flowed together in theories that bore the imprints of original minds, no longer bound within the limits of Han and T'ang philosophy. The result was a new outlook.

The most prominent of these Sung natural philosophers was Chu Hsi 朱熹. Utilizing the views of Chou Tun-i 周敦頤 and Ch'eng I 程頤, he made *li* 理 (a principle of organization) and *ch'i* the basis of a system. According to his cosmogony, in the beginning of Heaven and Earth there was only a unitary *ch'i*, revolving incessantly. As the speed of rotation increased, much impure residue from the *ch'i* gathered at the center. Since it could not escape, the Earth was formed there. The

[19] *Lun heng* 論衡, ch. 14.3, tr. Alfred Forke, *Lun-Hêng* (2d ed., New York, 1962), I, 119, modified. Note that Forke translates "*ch'i*" sometimes as "air" and sometimes as "fluid."

pure and clean *ch'i* on the periphery became the sky and its luminaries. This Heaven revolved about the motionless and central earth. The two formed an eternally immutable system of motion and quiescence corresponding to that of yin and yang. This idea of the separation of pure and buoyant *ch'i* to become Heaven, and of heavy and turbid *ch'i* to become Earth, goes back to the *Huai nan tzu,* but Chu Hsi made it explicitly centrifugal. From his unitary, bright, and supernal *ch'i* the dual and eternal *ch'i* of Heaven and Earth arose and were followed by the emergence of yin and yang. This schema is reminiscent of Lieh-tzu's stages. As the *ch'i* of yin and yang took on palpable character, the *ch'i* of the Five Elements, which represent the level of substance, appeared, and from them the material things of experience.

Again Chu Hsi took from ancient literature connected with shamanism the idea of a substantial ninefold sky, and turned it into the concept of a "nine-layer heaven" (*chiu ch'ung t'ien* 九重天). The nine strata of Heaven are formed by differences in the velocity of rotation of the ethereal *ch'i* that fills the universe. Where it is near the earth the *ch'i* is viscous and the rotation slow. Further outward, the *ch'i* becomes light and the rotation quick. The ninth stratum is infinite. The Heavens float on water, which also flows around and under the earth. The *ch'i* directly contiguous to the central earth does not rotate.

At the basis of this universal totality is the concept of Grand Polarity, derived from Chou Tun-i's (1017–1073) identification of the cosmogonic stage of nondifferentiation (*wu chi* 無極, literally, "nonpolarity" or, according to some commentaries, "the polar maximum of nonbeing") with the stage of differentiation which gives rise to cycles of change governed by complementary opposites (Grand Polarity or *t'ai chi* 太極).[20] This Grand Polarity is thus the underlying rhythm of change in the phenomenal world. In Chu Hsi's thought it became a primary principle of organization (*li*): "The Grand Polarity is the pattern-principle of Heaven and Earth and the myriad phe-

[20] "*T'ai chi*" is often rendered more literalistically as "Supreme Ultimate."

nomena. If one is speaking of Heaven and Earth, the Grand Polarity is located within Heaven and Earth. If one is speaking of the myriad phenomena, the Grand Polarity is located within the myriad phenomena."[21] What makes possible the existence of Heaven and Earth and the myriad phenomena— envisioned cosmogonically—is the principle of *ch'i*. What provides an opportunity, from the human point of view, to perceive and comprehend their existence are the fundamental principles of Grand Polarity and *li*. Chu Hsi saw a gradual transition from the ordering of *ch'i* to form Heaven and Earth to an ultimate recurrence of undifferentiated chaos. This oscillatory cosmogony, in which chaos and differentiation alternate age after age (somewhat analogous to Empedocles' reigns of Love and Strife), goes back to Shao Yung 邵雍 (1011– 1077). Shao had even attempted to put his cycles on a quantitative basis. He asserted that the great period of alternation, the Epoch Cycle (*yuan* 元), amounted to 129,000 years. It was composed of twelve Conjunction Cycles (*hui* 會); each of these was thirty Cycles (*yun* 運), each Cycle was twelve Generations (*shih* 世), each of which was thirty years. The first two terms are borrowed from the language of astronomy, but it is obvious enough that the scheme is analogical, based on a double parallel with the twelve Chinese hours in a day and the thirty days in a round month. The overall Epoch Cycle, which repeats endlessly, is divided into finite steps each of which Shao marked with one of the subordinate cycles.

Shao Yung accepted the Five Elements as a concept through which to understand physical change, but he put prior to them another set of four—Water, Fire, Earth, and Mineral, to which corresponded sun, moon, stars, and planets respectively. Since these four hypostases represented an earlier stage in Shao's cosmogony than the Five Elements (linked to the myriad phenomena), he applied to the former the Taoist term "anterior to Heaven" (*hsien t'ien* 先天) and called the latter "posterior to Heaven" (*hou t'ien* 後天). He saw the Five

[21] *Chu tzu yü lu* 朱子語錄 (Recorded discourses of Chu Hsi), ch. 1. Note that the myriad phenomena are ontologically posterior to Heaven and Earth.

Elements as the functional aspect of his fourfold principle. Shao's unique attempt to define the universe numerologically was very much in the style of Han mathematical astronomy, but the fourfold principle was his own application of the yin-yang dichotomy, and his cycles were analogical rather than based directly on observed celestial periods. Shao also attempted to work into his schema the appearance of man in the world and the movement of human history from the legendary sage emperors to the decline of his own time. In this sense he shared with his predecessors the goal of deriving the natural order and the political order from the same fundamental conceptions.[22]

Comparable in influence to the ideas of Chu Hsi on man and nature were those of his contemporary Lu Chiu-yuan 陸九淵 (also known as Lu Hsiang-shan 象山, 1139–1193) as completed and developed by Wang Shou-jen 王守仁 (or Wang Yang-ming 陽明, 1472–1529). Chu Hsi's li-ch'i theory of the universe stressed the role in individual development of knowledge and of investigation of the world, and did not give meditation and contemplation a major role. Lu put the latter first, for he was convinced that the truth is to be discovered within the mind rather than within the phenomenal world: "The universe is my mind, and my mind the universe."[23] This view, derived from the Buddhist "mind-only" (wei hsin 唯心) doctrine, was incapable of fostering theories of nature or understanding of the physical universe. Wang Yang-ming, in his quest for a unity of knowledge and action, reduced the principles of li and ch'i to an idealistic unity. His "innate knowledge" (liang chih 良知) was the pattern-principle of the natural order seen as immanent in the mind of man. This was the foundation of moral consciousness and action, since the same pattern-principle was inherent in the world upon which one acted.

But it was Chu Hsi's thought, considered authoritative for

[22] *Huang chi ching shih shu* 皇極經世書 (Supreme principles for governing the world), "Kuan wu 觀物."
[23] *Hsiang-shan hsien-sheng wen chi* 象山先生文集 (Collected prose of Lu Chiu-yuan), ch. 22.

purposes of the state examination system, which had the most influence in late traditional China, and much impact even in Japan.

Let us now return to the concept of the universal Way, embracing both nature and man, in religious Taoism. At the beginning of the fifth century the Taoist T'ao Hung-ching obtained Imperial patronage for his Mao-shan sect in south China, and a generation later K'ou Ch'ien-chih procured state sponsorship from the Northern Wei court. The religious Taoists were propagating a complete system of thought, in which Lao-tzu was not only considered the first philosophical exponent of the Way but had been deified and worshipped as *T'ai-shang lao chün* 太上老君, an aspect of the ineffable Tao. The canonical works of Taoism, as voluminous as the Buddhist Tripitakas and many times the length of the Confucian classics, were being compiled. Hu Ying-lin 胡應麟 (1551–1602) classified the contents of the fifteenth-century Taoist Patrology under five headings:

1. Purification and ataraxy (*ch'ing-ching* 清淨). Hu saw this as the meditative and ritualistic tradition described on the philosophical plane in the *Lao-tzu* and *Chuang-tzu* as the doctrine of ataraxy and noninterference (*ch'ing ching wu wei* 無為).

2. Alchemy and related disciplines, including physiological, meditative, and sexual techniques that frequently adapted the technical terminology of alchemy (*lien-yang* 煉養).

3. Dietary disciplines, including the ingestion of certain natural substances that conferred immortality (*fu-shih* 服食).

4. Amulets and other magical and apotropaic practices (*fu-lu* 符籙).

5. Liturgy and ceremonial (*ching-tien k'o-chiao* 經典科教).

Among the best-known thinkers who contributed to the ideology of religious Taoism was Ko Hung 葛洪 (or Pao p'u tzu 抱樸子, 283–343), also remembered as an official and physician (this manysidedness is characteristic of several of the great figures of religious Taoism). In discussing the principles of organization of Heaven and Earth, Ko Hung (following the lead of the *Lao-tzu*) called the generative origin of the natural order "the Mystery [*hsuan* 玄]." The Mystery lies at the bases of both conditioned and unconditioned existence (*yu wu* 有無). Its motion gives rise to *ch'i*, from which in turn Heaven and Earth are born:

The Chaos splits and the pure and turbid appear, [the former] rising and active, [the latter] descending and quiescent, and [become Heaven and Earth respectively]. Still I do not know why Heaven and Earth came to be as they are. The myriad phenomena, stimulated by the working of the *ch'i*, also spontaneously became one with Heaven and Earth. The only distinctions are of priority in formation and of scale in size.[24]

Ko's conception of *ch'i* is anything but fixed and stationary. As he put it, "the nature of Heaven and Earth is transformation (*pien-hua* 變化)."[25] This idea was in harmony with Confucian thought. The universe is always coming into being and passing away. The ultimate principle of change reduces to the activity of the gods, which can be known only through divination. This gnostic aspect was the true meaning of the *I ching*.

It is thus seen as essential that man confront transformation. Nature is transformation, and through the study of nature one reaches the more fundamental study of transformation itself.

"Gold which has been made by a transformation process is the seminal essence of the various ingredients."[26] In other words, alchemical gold is the unitary *ch'i* which underlies the phenomena of transformation, embodied in the outcome of the process. Thus for Ko Hung alchemical gold is far superior to the gold produced in nature and familiar to man. The material embodiment of this fundamental *ch'i* is gold, which itself is neither transformed nor born. Gold becomes a symbol of the perpetual and unchanging. It is contrasted with cinnabar (HgS), a symbol of change. When cinnabar is heated it becomes mercury, and when mercury is left alone it will revert to cinnabar. In other words, it exhibits an oscillatory or cyclic character. From the operation of these two modes, the unchanging and the transforming, it becomes possible for cyclic

24 *Pao p'u tzu nei p'ien* 抱樸子內篇 ("Inner chapters" of Pao p'u tzu), ch. 7.
25 Ibid., ch. 16.
26 Ibid.

phenomena to arise. In order to extract the seminal essences of various ingredients, alchemists tried all sorts of distillation techniques. By the use of an elixir which made use of the unchanging and constant nature of gold, the possibility of constancy—of attaining perpetual life or immortality—was opened to man.

As Ko Hung put it, men are endowed with varying amounts of *ch'i*, and the differences determine their respective fates.[27] Thus Taoists must devote care to techniques for augmenting their vital *ch'i*, and must take care not to waste it. The highest means of augmenting the *ch'i* is the Gold Elixir (*chin tan* 金丹), which incorporates an unchanging nature. Another method for eternally preserving man's *ch'i* is breath control (*hsing ch'i* 行氣), an augmentation of the *ch'i* by a kind of deep breathing. The "art of nurturing the vitality" (*yang sheng chih shu* 養生之術) guaranteed to adepts the possibility that by human endeavor one could manipulate the *ch'i* which underlies the myriad phenomena. This trend is significant for man's endeavor with respect to nature, but it is not in the main current of thought about human nature.

The Chinese view of nature was also expressed quintessentially in art, especially in painting. The canonic statement of the principles of painting was made in the sixth century by Hsieh Ho 謝赫:

1. the rhythm[28] of *ch'i* pulsating with life
2. use of the brush so as to structure the strokes correctly
3. form corresponding to the phenomena portrayed
4. distribution of colors according to the appropriate categories
5. composition and hierarchical subordination
6. transmission of classical models by copying.[29]

---

27 Ibid., ch. 13.
28 The art historians' conventional translation of "*ch'i yun* 氣韻" is "spirit resonance," but *ch'i* is an energetic concept and "*yun*" (written both 韻 and 運) is not a static quality but the temporal rhythm characteristic of the cyclic Tao of nature and life. "Spirit resonance" expresses part of the connotation of "*ch'i yun*," but it is not a translation.
29 *Ku hua p'in lu* 古畫品錄 (On the quality of ancient paintings).

Here too *ch'i* is given first place, for in no true painting is this resonance lacking. The matter of semblance is treated in the third and fourth principles with reference to form and color, but the priority was given to the structural expression of *ch'i*.

Early landscape painting was concerned with Taoist nature mysticism, developing gradually a concern with the depiction of nature in its own right. For instance, the earliest extant detailed description of a great landscape painting by the artist himself is Ku K'ai-chih's 顧愷之 (344–406) essay on a depiction of the Cloud Belvedere Mountain (Yun-t'ai shan 雲臺山). The focus of this elaborate composition was the Taoist Celestial Master Chang Tao-ling in meditation with his disciples at the edge of an abyss in a majestic mountain setting.[30] The conception of *ch'i* rhythm, formed out of the criticism of portraits, also became the chief desideratum in landscape painting. The secondary place given to mere simulation of form is also reflected in the six essentials of landscape painting formulated in an essay attributed to Ching Hao 荊浩 (late tenth century). They are in reality three pairs:

ch'i—cosmic rhythm
thought—motif
brushwork—ink quality.[31]

The relation between immanent *ch'i* and outer shape is likened to that of a seed and the blossom which grows from it. As Teng Ch'un 鄧椿 put it in 1167,

The range of painting is very wide; it comprises everything between heaven and earth. By revolving their thoughts and preparing the brush the painters can represent the characteristics of everything, but there is only one method by which it can be done thoroughly and exhaustively. Which is that? It is called "being in touch with divinity." People think that men alone have spirit; they do not realize that everything is divine.[32]

[30] In *Li tai ming hua chi* 歷代名畫記 (Record of great paintings in successive eras), ch. 5.

[31] *Pi fa chi* 筆法記 (On brushwork).

[32] *Hua chi* 畫繼 (The tradition of painting), tr. Oswald Siren, *The Chinese on the Art of Painting* (Peiping, 1936), pp. 88–89, modified.

This lies at the basis of Heaven and Earth and all their phenomena, for it is said that divinity (*shen* 神) and *ch'i* are almost equivalent in sense.

Thus to paint a landscape implies extracting from nature its laws in order to depict what lies at nature's base. Because painting expresses the conceptions of the painter, the boundless space suggested in Chinese landscape painting has nothing in common with the lifeless vacuum that Western thinkers following Aristotle thought must be abhorrent to Nature. This Chinese space was the plenum of *ch'i*, continuous and unitary with the *ch'i* of mountain and stream.

The Sung painters also brought out the rational pattern of the cosmos in their work. Even in things of indefinite form—waves and mist—they knew there lies a constant principle of organization. "In Heaven and Earth are many things. If they contain some rationality they will not be random. No matter how many things there be, if one begins from the beginning there will be no Chaos. This is because there is a principle of organization connecting all things." That is how Han Cho 韓拙 put it in 1121.[33] The Sung philosophers' use of the principle of organization (*li*) and the notion of *ch'i* to explain the world was adapted in order to comprehend that human activity which was devoted to depicting Heaven, Earth, and all they contain. Landscape paintings precisely express the Chinese style in natural philosophy. As T'ang Hou 湯垕 said in the Yuan,

Landscape painting is the essence of the shaping powers of Nature. Thus through the vicissitudes of yin and yang—weather, time, and climate—the charm of inexhaustible transformation is unfailingly visible. If you yourself do not possess that grand wavelike vastness of mountain and valley within your heart and mind, you will be unable to capture it with ease in your painting.[34]

Only one who has the divine *ch'i* of Nature concealed in his mind can understand and express Nature's divine *ch'i*. The immanent law of Nature, incorporating inexhaustible transformation, is thus the immanent law of man.

[33] *Shan shui ch'un ch'üan chi* 山水純全集 (On landscape painting).
[34] *Hua chien* 畫鑒 (Mirror of painting).

5

Chinese Astronomy:
Development and Limiting Factors

Kiyosi Yabuuti

The basic pattern of Chinese astronomy was formulated in most of its essentials during the Former Han dynasty. Thereafter, in the course of its slow evolution until the nineteenth century, we find no revolutionary modification of a magnitude commensurable with that of the change from geostatic to heliostatic theory beginning in sixteenth-century Europe.

In China, as elsewhere, astronomy can be traced back to the dawn of civilization. During the Yin, the dynasty that grew out of the Urban Revolution, astronomy began characteristically as calendrical science. The period of the waxing and waning of the moon (the synodic month, 29.53 days) was expressed in terms of a "great month" of thirty days and a "small month" of twenty-nine days. In order to bring consonance between tropical years (365.24 days) and synodic months, the length of a calendar year was taken as twelve calendar months (about 354 days) for an ordinary year and thirteen months (about 383 days) for a leap year, which occurred every two or three years. A major technical problem had to do with the proper spacing of the intercalary months. In very early times they were added when necessary to keep the calendar in step with the sky, on a more or less *ad hoc* basis, but later calendars about which we have adequate information used intercalation schemas such as the Rule Cycle 章 (called the Metonic Cycle in the West, where it was also used), according to which seven months were added at roughly equal intervals over a period of nineteen years. But cycles of this sort, which make possible the compilation of calendars that will tally with the celestial phenomena over long periods, do not go back to the Yin.[1] There is no

---

[1] I do not believe that Professor Tung Tso-pin has succeeded in proving such an early origin. See Tung Tso-pin 董作賓, *Yin li-p'u* 殷曆譜 (Chronology

evidence that the Chinese borrowed their calendrical techniques from Europe. During the long interval between the Yin and the Warring States period, these numerical relations were gradually worked out.[2] The Chinese, a pragmatic people, were not much inferior to the Greeks in compiling calendars for practical use.

The Warring States period, when the Quarter Day calendrical system 四分暦 was devised, was one of the most creative periods in the history of Chinese civilization. Along with the improvement of the calendar, much effort was given to observations of celestial bodies. In particular, the motion of Jupiter over a cycle of about twelve years gave rise to the notion of geographical association—or, as the Chinese called it, "field allocation 分野." The circumference of the sky along the equator was divided into twelve equal parts, or "stations 次," and each station was understood to correspond to one or two feudal states. Observing the position of Jupiter among its stations, the astrologer obtained an interpretation and applied it to the corresponding feudal state or its lord.[3] This division resembles the Western zodiac, although the latter is divided along the ecliptic. Also used was a system of twenty-eight unequal equatorial divisions, the "lunar mansions 宿," which are nearly the same as the Indian *nakshatra*. Although in modern Chinese and Japanese the word "*t'ien-wen* 天文" refers to the exact science of astronomy, in ancient times it meant the pattern of the sky as interpreted by astrology and was extended to that art itself. As the Book of Changes put it, "the heavens manifest good and evil signs through the celestial phenomena."

---

of the Yin; Nan-ch'i, Szechuan, 1945), and my *Chūgoku no temmon rekihō* 中國の天文暦法 (Chinese astronomy and calendrical science; Tokyo, 1969). Additional bibliography is available in Tung Tso-pin, *Fifty Years of Studies in Oracle Inscriptions* (Tokyo, 1964 [?]).

[2] The late Professor S. Shinjo argued that the Chinese discovery of the 19-year and 76-year intercalation cycles predated Meton (432 B.C.) and Callippus (334 B.C.), after which they were named in the West. See Shinjō Shinzō 新城新藏, *Tōyō temmongakushi kenkyū* 東洋天文學史研究 (Researches in the history of Oriental astronomy; Tokyo, 1928), p. 30.

[3] Examples of this kind of divination are found in pre-Han classics, notably the *Tso chuan* 左傳 and *Kuo yü* 國語.

Celestial portents were not merely natural phenomena, but expressions of the will of Heaven communicated to the ruler as admonitions. According to the Chinese theory of monarchy, the supreme ruler was the Son of Heaven, and through him the celestial will was to be transmitted as the basis of social order. Though the Chinese Heaven is neither a creator nor a god in the theological sense—later, seen more philosophically, it *was* the cosmos or natural order itself—it provided criteria for moral and political conduct and thus occupied a crucial position in Chinese political ideology. To supervise the heavenly ritual was the ruler's privilege as well as his duty, for it was an essential service which only he could perform on behalf of his subjects.

There are two sorts of celestial phenomena. One was cyclical in a simple way, and its regularity or periodicity could be discovered with relative ease; the other could not be predicted by human effort, but only observed. The former were systematized within the framework of calendrical science, while the latter became the objects of astrological interpretation. Since they were complementary, they were equally important to the Chinese administrators. In official dynastic histories from the *Shih chi* 史記 (ca. 90 B.C.) on, two chapters have usually been devoted to these disciplines, a "Treatise on Harmonics and Calendrical Astronomy 律曆志" and a "Treatise on Astrology 天文志."[4]

Unlike medieval Europe, where religion dominated every aspect of society, in China political values took precedence in all matters after the political unification of the Ch'in and Han. A bureaucracy, with the Emperor at its top, was firmly established; the despotic character of the monarchy was able to exert itself more powerfully as time went on. Even astronomy was not too abstract to escape being decisively formed by its

---

[4] T'ang sources tell us that when predicted eclipses did not occur, officials congratulated the Emperor for having suppressed them by the strength of his virtue. See Yabuuti (ed.), *Chūgoku chūsei kagaku gijutsushi no kenkyū* 中國中世科學技術史の研究 (Studies in the history of science and technology in medieval China; Tokyo, 1963), p. 19.

political applications. The astrological treatise of the *Shih chi* was entitled "Treatise on the Celestial Offices 天官書"; the names of the constellations and those of the bureaux of the imperial administration corresponded. This was, of course, a reflection of the correspondence between earth and sky— between the natural and political orders—which lay at the basis of Chinese astrology, but here we see it given articulation by the special character of government. Astrological interpretation was purely judicial in character, paralleling Babylonian practice. Its main function was to prognosticate the fates of rulers and states, not like horoscopic astrology, which developed in the Hellenistic world to foretell the fortunes of individuals. The latter was unknown in China until the end of the T'ang, when it was imported from the West, but even then it never replaced or noticeably altered native forms of private divination.

The supremacy of politics also shaped calendrical astronomy. Today's calendars deal only with years, months, and days, a reflection of their original function, the prediction of seasonal events and phases of the moon. In traditional China, however, the corresponding concept had much wider implications. The object of study of *li-fa* 曆法, calendrical science, was not only the composition of calendars for daily use, but also the calculation of all other known celestial regularities, for instance eclipse prediction and computation of planetary positions. The calendrical treatises, especially those of the later histories, were in many respects equivalent to the modern *Astronomical Ephemeris* or to Simon Newcomb's astronomical tables. Whereas the calendrical treatise of the *Shih chi* described the method of the Quarter Day calendrical system, limited to intercalation of days and months, the "Treatise on Harmonics and Calendrical Astronomy" of each dynastic history since that of the Former Han treated all regular celestial phenomena (that is, those which could be computed at the time). The breadth of the Chinese ephemerides reflected the grave concern of Chinese rulers constantly to expand the demonstrable order of the sky, while reducing the irregular and ominous. The parallel with the ruler's responsibility in the political realm is obvious.

The Astronomical Bureau was tightly incorporated within the bureaucratic framework under the despotic rulers of the Han dynasty. The compiler of the *Shih chi*, Ssu-ma Ch'ien, was made head of the bureau, at that time called Office of the Grand Historian 太史, because the functions of archivist and recorder of omen observations were not yet differentiated. Afterwards this post became specialized, and its incumbent can justly be called Astronomer-Royal. Though the name of the office, the title of its director, and its precise functions changed from time to time through the long history of China, the Astronomical Bureau continued to exist without radical alteration for two thousand years, until the demise of the Ch'ing dynasty in the twentieth century. Its main functions remained research in calendrical science, compilation and distribution of calendars and ephemerides, and unremitting observation of astrological omens. Maintenance of water clocks and other timekeeping services, as well as hemerology (the designation of calendar days as lucky or unlucky in general or for certain kinds of activity), were also assigned to the Bureau.

In medieval Europe the Pope established the legitimacy of the Holy Roman Emperor, but the former had no counterpart in China. The founder of a new Chinese dynasty had to demonstrate the transition of legitimacy, after his victory, partly by his performance of certain imperial rituals and partly by reforming certain institutions which were closely related to the imperial charisma. One such practice was "correcting the beginnings of years and months 改正朔"—that is, calendar reform. From the beginning of the Han this had been a major issue, and finally it was accomplished officially for the first time in the first year of the Grand Inception 太初 era (104 B.C.) under Emperor Wu. This new system was named the Grand Inception calendar, and was further revised as the Triple Concordance 三統 calendar at the end of the Former Han by the scholar-official Liu Hsin 劉歆. The system of the Triple Concordance calendar became the basis of the calendrical treatise of the History of the Former Han (*Han shu* 漢書).

Let us now look more closely at the basic pattern of Chinese astronomy as essentially formulated during the Han dynasty.

First of all, astronomy was defined as a field of research. Celestial observations for judicial astrology and the computation of exhaustive ephemerides were major responsibilities of the official astronomers. From the Han to the Six Dynasties period, a variety of cosmological theories were proposed and debated among scholars, but these arguments were tacitly rejected by professional astronomers from the T'ang on. The methods of research in calendrical science also took a rather narrow form. Chinese calendrical science was, at least in principle, applicable to all periodic celestial phenomena; but unlike Greek astronomy, which employed geocentric models based on a well-developed Euclidean geometry, the Chinese, who were able arithmeticians and algebraists, succeeded in finding regularities in celestial phenomena without recourse to geometrical models. Even before the Grand Inception system the Chinese knew the regularity of the tropical year and the synodic month, and what they proceeded to look for in the heavens were indications of analogous simple periodicities. In the Triple Concordance system, we notice their discovery of regularities, such as the 135-month period of lunar eclipse recurrence and the synodic periods of the planets.[5] Reliance on simple periodicities of this kind alone could not lead to success in predicting the whole range of phenomena. Although astronomers moved in the direction of less linear computation methods continually from the Han on, the fundamentally metaphysical assumption that all phenomena could be expressed in terms of simple periodicities long survived its utility.

The Greek geometrical model implies real movements of the heavenly bodies in three-dimensional space. The work of Greek

[5] This period is quite distinct from the 223-month eclipse cycle known to Pliny in the West and must have been discovered independently. On the other hand, the adoption of armillary spheres and equatorial coordinates suggests some influence from the West around the first century B.C. See *Chūgoku no temmon rekihō*, p. 13, and N. Sivin, "Cosmos and Computation in Early Chinese Mathematical Astronomy," *T'oung Pao*, 1969, *55:* 1–73. An overview of modern Japanese research on Chinese astronomy as a whole is provided in Shigeru Nakayama, *A History of Japanese Astronomy. Chinese Background and Western Impact* (Cambridge, Mass., 1969).

astronomers had consequences for the physical understanding of celestial phenomena, and thus eventually for cosmology. The algebraic techniques of the Chinese official astronomers were not tied to any physical hypothesis, and none was offered explicitly in the calendrical treatises. Although their object was restricted to the prediction of celestial phenomena, it is not realistic to conclude that they did not personally share one or another prevalent philosophical conception about the real spatial character of celestial motions.[6] But these conceptions were not applied to their technical work, and they did not have the habit, especially after the Han, of dwelling on the theoretical basis of their computations. This approach, whose metaphysical assumptions lay in the realm of time cycles rather than spatial relations, was perpetuated as the paradigmatic style of Chinese astronomical research.

The control of astronomical research by the bureaucracy at least assured continuity through the turbulent centuries. After the overthrow of a dynasty, the need to establish the legitimacy of the successor meant that the Astronomical Bureau would soon be restored and routine work not seriously interrupted. Although the political implications of calendrical science were weakened after the Chin period, the patronage of the state facilitated slow but steady accumulation of voluminous observational records and improvement of astronomical tables. Although the founder of a new dynasty tended to be sympathetic toward revolutionary ideas and major institutional alterations, his successors were generally more concerned for the status quo, and a conservative attitude prevailed. Chinese astronomers were on the whole bureaucrats first, following established routines to discharge established responsibilities, and only secondarily researchers. It is natural enough that their conservatism inhibited concern with new problems or

[6] One can infer from the much improved eclipse computation procedure of the T'ang that astronomers were seeing the problem as one of the moon's passage through the earth's shadow. See Yabuuti, *Zuitō rekihōshi no kenkyū* 隋唐曆法史の研究 (Researches in the history of calendrical science during the Sui and T'ang periods; Tokyo, 1944), p. 104.

new approaches to old ones. Of course there were many significant exceptions over two millennia. The inequality of the lunar motion was noticed in the Later Han and that of the sun at the end of the Six Dynasties era. Computation technique was remarkably improved at the beginning of the Sui, when interpolation formulae were devised. A number of other innovations were incorporated in the grand synthesis of Kuo Shou-ching 郭守敬, the culmination of traditional astronomy at the beginning of the Yuan. But these efforts were still almost exclusively devoted to the improvement of the ephemeris, for the focus of Chinese astronomy remained on the field of calendrical computation.

The civil service examination system played an important role in shaping the Chinese bureaucratic regime. After its adoption in the Sui, it strengthened the despotic power of the rulers. At the end of the Ch'ing, many politicians and scholars argued that this system of merit examinations had been a major obstacle to the development of science. They pointed out that it required mastering only the Confucian classics and literary works and thus offered no recognition to those who studied science or technology. The chief merit of the examination system lay in its egalitarianism, in its encouragement of social mobility, as opposed to the strict hereditary system of feudal Japan. The examinations were public, open to everyone, and held out good careers to those who survived the stiff competition. They did much to prevent dissatisfaction and frustration among men of talent and hence deserve some credit for the long survival of the bureaucratic imperial system.

Although science and technology were not means to a grand career in this system, there were cases aplenty in which high officials maintained and put to good use amateur interests in science and technology, especially astronomy. Calendrical problems were, in principle, the business of specialists in the Astronomical Bureau, but high officials outside it, and even commoners, often contributed their opinions, and were listened to. Most members of the bureau itself were minor officials, technicians, who could not look forward to being promoted to the top.[7] Without much incentive to show initiative, they tended to stick to routine, and—especially from the T'ang on, when much of the computational work was done by foreigners armed with the superior techniques of Indian or Islamic astronomy—often did not take their duties seriously.[8] On the whole, the examination system was effective in selecting talented candidates for purely administrative posts, but it did not serve the advancement of science.

The feudal system of Tokugawa Japan, despite certain formal similarities in court organization, was utterly different in spirit. Hereditary succession to office was strictly maintained, although by the end of the Edo period it had been much compromised. Regardless of what talent a man had, there was no prospect of promotion to higher rank. Those whose frustration had no conventional outlet visited Nagasaki—the town to which Dutch merchants were restricted, and thus the only window open to Europe at the time—and concentrated on absorbing European science. This was one of the reasons "Dutch learning" prevailed and ultimately influenced the early modernization policy of the Meiji Restoration in the late nineteenth century.

In spite of the predominance of stability-oriented political values, we find, looking back at the past, that periods of political disturbance did not inhibit scholarly activity. In the Warring States period, for instance, the "Hundred Schools" of philosophy, which defined the perennial problems and approaches of Chinese civilization, flourished. Remarkable astronomical progress was made in the transitional period from

---

[7] Specialists in calendrical science and astrology were in general not eligible for transfer to other government offices.

[8] In the Northern Sung there were two observatories, one inside and one outside the palace. Whenever anomalies appeared in the night sky, the astronomers of the inner observatory were supposed to report them immediately; the next morning, upon the opening of the palace gates, the reports were to be checked against the records of the outside observatory. In practice, the two staffs simply cooked the records, depending upon their collusion to avoid discovery. See Joseph Needham, *Science and Civilisation in China* (7 vols. projected; Cambridge, England, 1954–), III, 191–192.

the Six Dynasties to the Sui. The medical theories of Li Kao 李杲 and Chu Chen-heng 朱震亨, the greatest achievements of the algebraic tradition, and many preliminaries to the Yuan astronomical synthesis were all made in the short interval of dynastic change from the Chin to the Yuan. It must be pointed out, however, that political disturbances sometimes cut short new lines of scientific research, as was the case in the Northern Sung. From the beginning of the Sung dynasty, scholar-officials gave special attention to collecting and collating old scientific treatises, and there was much exploration of and technical research on new problems. These trends were cut short by the invasion of the Chin Tartars (completed by 1126). The academic tradition survived, much attenuated, in North China under the Chin occupation, while a more popular culture, based in the rising merchant class, flourished in central and southern China under the Southern Sung. This bifurcation continued into the Yuan, although the old academic tradition gradually disappeared in North China as Central China gained economic importance and intellectual influence.[9] Thus academic learning did not become one of the strengths of the Ming period.

Joseph Needham has pointed out in his survey of Chinese science that before the fifteenth century there was not much disparity in the level of scientific achievement between China and the West. Chinese inventions contributed to the awakening of European civilization in the Renaissance. From the beginning of the sixteenth century the Portuguese were in contact with China. Within the century Jesuit missionaries had entered the mainland of China. Through their efforts the superiority of the West in the field of mathematical science was recognized by some Chinese, a few of them highly placed. Missionaries translated a number of European scientific treatises into Chinese. Sympathetic court officials of high rank, who believed in the value of European astronomy, proposed a calendar

[9] The scientific achievements of the Northern and Southern Sung have been discussed in Yabuuti (ed.), *Sōgen jidai no kagaku gijutsu shi* 宋元時代の科學技術史 (History of science and technology in the Sung and Yuan; Kyoto, 1967), pp. 81–86.

reform based on Western methods. Since the Chinese had been depending upon foreign astronomers for some time, and the Jesuit techniques were shown to be the best available, the proposal was accepted. A preliminary compilation of considerable bulk, a veritable encyclopedia of European astronomy (the *Ch'ung-chen li shu* 崇禎曆書) was completed by Chinese and missionaries working in collaboration. The Ming house was shortly overthrown, and Hsu Kuang-ch'i 徐光啓, the major Chinese promoter, died prematurely, but the reform was promulgated in 1645 under the Ch'ing. This gigantic project succeeded partly because of Hsu's power to overcome opponents of "barbarian" innovations and partly because of the fact that he maintained the basic framework of the traditional Chinese calendar, with European computational techniques and astronomical parameters injected when appropriate. During the Jesuit period, the partial acceptance of European astronomy did not necessitate a revolutionary shift in Chinese calendrical science.

Besides the limitations imposed by the institutional status of Chinese astronomy, the role of geographic setting in encouraging a homeostatic orientation of society is also reflected in astronomy—most strongly in this discipline's comparative lack of openness to foreign influence.[10] The natural barriers on all sides of China made her not often vulnerable to invasion, but the lack of a diversity of high cultures within these barriers encouraged the provincial view that China was the only true civilization in the world. Foreign conquests and occupations, always by peoples whom the Chinese could consider barbarian, brought little cultural influence of a kind the Chinese valued or were ready to acknowledge. They were continuously in contact with India, and later Islam, but of all these cultures offered, only Buddhism had a major and lasting influence. Indian astronomical techniques reached China during the T'ang period, and those of Islam in the Yuan and Ming, but

[10] The word "homeostatic" has been used by Dr. Needham in place of "stagnant." See "Science and Society in East and West," in (among other places) *Centaurus*, 1964, *10*: 189.

not in a form which made them appear superior to the native methods, and the foreign specialists usually worked in isolation in the court. The Jesuit astronomy which was introduced during the Ming and Ch'ing was basically characteristic of Europe *before* the Copernican Revolution. In 1650 it may have resembled that taught by conservative astronomers in Europe, but by 1700 it was ludicrously obsolete.[11] Even so, with spherical geometry, trigonometry, and logarithms, it was from the beginning far superior to the Chinese art. That it was very imperfectly transmitted in the Jesuits' Chinese writings is partly because the missionaries were not promoting it for scientific ends. They had no interest in fundamentally reforming Chinese astronomy. Their astronomical activity was merely an entrée to the court Astronomical Bureau, which they meant to use as a center for evangelizing China from the top down. The volume of their astronomical writing fluctuated inversely with the security of their position in Peking.

In Europe, in a territory more confined that that of mainland China, many peoples, differing in language and tradition, intermingled to form a European civilization that was bound to be supranational in important respects. Non-European ideas and technics were also assimilated in the making of modern Europe; the roles of gunpowder, papermaking, printing, and the compass were central to Europe's consolidation and expansion in the transition from the Middle Ages to modern times. A climate inconducive to science in one European country could be balanced by favorable circumstances in another. When the pressures of the Church aborted the scientific eminence of Italy in Galileo's time, and incessant war and religious turmoil in Germany made fruitful lifetimes of work impossible for many men less impervious than Kepler, there ensued a natural shift in the center of gravity of research to England and France. One need not look for racial factors in the rise of modern science in the West rather than the East.

[11] See N. Sivin, "Copernicus in China," forthcoming in *Studia Copernicana* (1973).

Because China remained one homogeneous unit intellectually isolated from other major civilizations, it was protected from the sorts of exterior challenge which would have forced radical change in patterns of society and of scientific research.

6

A Systematic Approach to the Mohist Optics (ca. 300 B. C.)

A. C. Graham and Nathan Sivin

The Mohist school was the earliest of the philosophical rivals of Confucianism, and lasted from its foundation by Mo-tzu late in the fifth century until at least the third century B.C. It was an organized school, with exactly ten doctrines defended in turn in ten triads of chapters (*Mo-tzu,* ch. 8–37), for example, "promotion of worth," "universal love," "rejection of aggression," and "rejection of fatalism."[1] The tenets of the school condemned offensive but not defensive war, and its members included specialists in military fortification and engineering, which are the subjects of the military chapters of *Mo-tzu.*[2] This and other signs of concern with the crafts and of low social origins suggest that the school was rooted in the artisan class.[3] The remarkable social fluidity of the late Warring States period makes such a genesis comprehensible. Its uniqueness in Chinese history does much to explain the lack of Mohist influence, despite great originality, on later Chinese philosophy.

The Mohist Canon (Mo-ching 墨經)
Two primary interests of the school underlie the remarkable document commonly called the Mohist Canon (composed of two chapters of Canons 經 [40 and 41] and two of Explanations of the Canons 經說 [42 and 43]): the sharpening of the tools of disputation for debate with rival schools, and the clarification of problems of physics raised by the craft of military engineer-

[1] *Mo-tzu,* 49.61–64. Our references to the *Mo-tzu,* unless otherwise noted, are to the Harvard-Yenching Institute Sinological Index Series text. Full references for this and other sources are given in the Bibliography following this chapter (p. 147).
[2] Ch. 52–71. The best available study of the military chapters is by Ts'en Chung-mien, who sees the scientific propositions being applied there.
[3] Watanabe, "Bokka no shūdan to sono shisō," pp. 1221–1223.

ing.[4] These two interests converged; the Canon contains, for example, a series of geometrical definitions that seem to be designed equally to establish a consistent technical terminology and to avoid paradoxes of space and time raised by the sophists of the fourth century B.C.

The Mohist Canon, which is followed by two more chapters on disputation, falls conveniently into five divisions:

A1–A75. Definitions of key words in ethics, government, disputation, and geometry, all loosely grouped in series. Those with most bearing on the scientific sections are the terms for kinds of change, motion, and rest (A44–A50), and the geometrical terms (A52–A69).

A76–A87. Multiple definitions of ambiguous words.

A88–B12. An examination of the methods and fallacies of different kinds of argument. This, with the further advanced and therefore presumably later *Hsiao ch'ü* 小取 (ch. 45) and the fragmentary *Ta ch'ü* 大取 (ch. 44), presents the Mohist semantics and proto-logic, which is as near to a true logic as ancient China arrived.

B13–B31. Propositions concerning the sciences, about space and time (B13–B15), optics (B16–B23), mechanics (B24–B29), and economics (B30, B31).

B32–B82. Solutions to controversial questions, many of them replies to fallacious claims of other philosophical schools. A number of the assertions refuted are easily identifiable as theses of Chuang-tzu 莊子, Hui Shih 惠施, Sung Hsing 宋鈃, and Kao-tzu 告子, all of the fourth century B.C. This suggests a date not long before or after 300.

In the later Mohist schools the habit of rational thinking, sharpened by debate with competing philosophical schools, combined with the curiosity of the builder or carpenter begin-

---

[4] We use the word "Canon" to refer to individual sections of the *ching*, "Explanation" for corresponding sections of the *ching shuo*, "Mohist Canon" for the *Mo-ching* (ch. 40–43 of the *Mo-tzu*) as a whole, and "Proposition" for an individual Canon and its associated Explanation considered together. We follow T'an Chien-fu's numbering of the propositions in the Mohist Canon.

ning to puzzle over the principles behind his craft. Anyone familiar with the history of science will be aware that this is an explosive combination which did not ignite until about A.D. 1600 in Western Europe, and that there is great interest in the infrequent episodes in earlier history when the two factors, normally isolated from each other at different social levels, entered into fruitful interaction. Among the scientific sections of the Mohist Canon, the mechanical propositions consider problems of balance, support, and free fall suggested by the design of steelyards, pulleys, ladders, and walls. The eight optical propositions have no such direct connection with technology, although it merely perpetuates an artificial distinction to ask whether the Mohist inquiries were guided by practical need or disinterested curiosity. Certainly the problems that excite the Mohist are how an object can cast two shadows, how its shadow can fall between itself and the sun, why an image is inverted in a concave mirror. It is interesting that one datum about the concave mirror which he does not adduce is its practical function in igniting tinder. No one else in China seriously asked the same sorts of question until Shen Kua 沈括 (1031–1095). He may well have been unaware of his Mohist predecessors, since the last period prior to the eighteenth century when the Canon was widely studied was about A.D. 300.[5]

The Canons and the Explanations
Since Canons and Explanations are grouped in separate chapters, the task of coupling them has engaged scholars ever since the study of the *Mo-tzu* revived in the eighteenth century. Fortunately most Explanations carry as heading the first word of the corresponding Canon, and in Part B (ch. 43) the Canons end with references to their Explanations in the form "Explained by 說在...." A Canon such as B20 ("The size of the shadow. Explained by: tilt and distance.")

[5] For a short survey of Chinese optics, see Needham, IV.1, 78–99. A document of Shen Kua on the concave mirror and the camera obscura is translated in Appendix B to this chapter (p. 145) below.

leaves no doubt that explanations in some form must be as old as the Canons themselves, and certainly both show every sign of belonging to a homogeneous tradition. But there is seldom the same exact verbal correspondence in the "explained by" formula as in the headings, which suggests that the formula refers to an earlier verbal exposition. Very probably the document as we know it accumulated gradually, with the Canons from the start written or at least standardized to be learned by heart, but with the Explanations for a long time oral and fluid.

Canons in the scientific part lay down controversial claims (B16, B18) or problems whose solutions are by no means obvious, followed by a summary indication of, or reference to, the principle on which a correct understanding is to be based ("explained by remaking, redoubling, point, etc."). The Explanation is in turn an expansion or illustration of this principle. The scientific Explanations tend to be fairly comprehensive, but in B16 and B17, as often elsewhere, one has the impression that knowledge of the problem is assumed and that the Mohist is making only what for him is the crucial point. A modern interpretation which merely makes sense is irrelevant unless it gives the Canon a point, shows what was puzzling people. As for the Explanations, they range from brief jottings to full expositions, but seldom or never stray far from the topic. One has no right, for example, to find the theme of multiple mirrors in the Explanation of B23 unless one can connect it with the Canon.

Problems of Comprehension and Translation
The Mohist Canon, as well as the dialectical and military chapters of *Mo-tzu*, present textual problems of extreme complexity. At the time when Alfred Forke translated the Mohist Canon as part of his German version of *Mo-tzu*, a great deal of it was unintelligible. Although progress has been made since, much of it will no doubt remain so. Translators into English have chosen to omit the Mohist Canon rather than write pages of nonsense, which is the price always paid so far by those who

have attempted integral translations of such texts as *Mo-tzu* and *Chuang-tzu*. One might wish, however, they had made it plainer that they have omitted what from several points of view are the most important chapters of the book. It is no coincidence that the most satisfactory translation of the whole *Mo-tzu* into any language is by a distinguished historian of science, Yabuuti Kiyosi.

The difficulties are of two kinds:

1. Since Canons and Explanations are printed without intervals and grouped in separate chapters, the work of fitting them to each other has occupied scholars ever since Pi Yuan's edition of 1783. Fortunately in the case of the optical part the task may be considered completed. Modern editors (T'an, Kao, Wu, and Liu) fully agree over the divisions, except that Wu makes the division between the Explanations of B21 and B22 two characters earlier than other scholars, choosing a different occurrence of the word *chien*, "mirror," to serve as the heading of B22. Comparison between Canons and Explanations also reveals a major dislocation of the former involving the optical sections, with B14–B20 and B21–B23 transposed. It has only recently been understood that one of the breaks was in the middle of B23, which can be reconstituted by joining two separated fragments.

2. There is an extraordinary profusion of rare or unknown graphs, without radicals or with unusual radicals, and words are often differently written in the same context. It would seem that the adaptation of ancient texts to later graphic usage was frustrated in the case of the Mohist Canon by inability of early editors or scribes to comprehend what it was saying. A point that does not seem to have been sufficiently noticed is that obscure graphs, when identified with philological rigor, turn out more often than not to be common words written in unfamiliar and sometimes corrupt forms. Fortunately the vocabulary tends to be simple, consistent, and free of idiom, with words used in their ordinary senses except when employed as words of art. Although scholars have done much work searching for rare words or senses of words in the *Shuo wen chieh tzu* 說文解

字 or *Fang yen* 方言 lexica, most of their labor has been wasted. Attempts to find rare functions of particles are even less justified; the syntax is not less but more consistent than in most pre-Han texts. It may be added that the extent of graphic corruption has been grossly exaggerated. The habit of trying conjectural emendation as a first instead of a last resort, seen at its worst in the edition of Kao Heng, has provided more distraction than aid.

In China, since the revival of interest in the corpus after over a millennium of obscurity by Pi Yuan near the end of the eighteenth century and Sun I-jang a century later, there have been many studies, philological and substantive, of the Canons, both separately and as part of the Mohist literature. As modern science became established in China, and the question of ancient Chinese scientific developments became central to the search of the less alienated scientists for their cultural roots, it was inevitable that they should bring to the Mohist Canon a sophisticated theoretical and practical knowledge of what physical realities the optical and mechanical propositions could have been about. The first to attempt a full-scale reconstruction of the scientific content was T'an Chieh-fu, an electrical engineer by training, in a fundamental study of the whole canon (1935).[6] The first physicist, Ch'ien Lin-chao, published five years later, and by now a substantial literature of this kind has accumulated. As part of his epochal *Science and Civilisation in China,* Joseph Needham has surveyed several important contributions, producing in the process better translations of the scientific propositions than Forke's and, equally important, establishing extensively for the first time the connections of the physical propositions with China's other contributions to scientific thought.

Despite the volume of recent work on the Mohist Canon, it was the need for a closer union of philology and physics which led to this collaboration. One of us (A.C.G.) has spent

[6] The first monograph on the optics of the Mohist Canon was a quarter-century earlier, by Feng Han-ch'u, but this work has not been available to us (nor was Yen Ling-feng, the *Mo-tzu* bibliographer, able to see it).

some years studying the proto-logic of the Mohist Canon and its textual and linguistic problems. The other (N.S.), led by a concern with understanding Chinese scientific thought from the inside and prompted by the important critical review of the literature by the physicist Hung Chen-huan, undertook a study of what experiments would have occurred most naturally to an ancient Chinese physicist if, unlike his modern annotators, he did not have access to a textbook of optics. We both felt that the major obstacle to carrying our comprehension of the documents as far as possible is the tendency of classicists who are unsure of their physical knowledge, and of scientists who have never had an opportunity to master the disciplines of philology, to relax their standards of rigor in areas where their confidence is least. Individuals have shown considerable strength in both fields—among others, one thinks of T'an Chieh-fu and Hung Chen-huan. But to arrive at a clear determination of what one can hope to comprehend given the present state of *Mo-ching* studies, and of what must be left for some unanticipated breakthrough, several methodological principles must be applied simultaneously:

1. All readings must conform to the language of the Mohist Canon as a whole, in point of syntax and diction. This is a coherent and consistent language, more or less the standard "third century B.C. dialect" of Karlgren. It does of course have enough peculiarities to deserve special study.

2. Emendation must be considered not an armchair sport but a last resort, to be undertaken only where the mechanism of corruption can be reconstructed and the plausibility of the particular emendation documented. With a text as patchy as that of the Mohist Canon, there are inevitably times when one is merely aware that something is wrong and has nothing to go on but taste and a general sense of style in suggesting a better reading. Here it is necessary to state explicitly that one is producing a reasoned speculation, not a solution.

3. Technical terms must be rendered consistently throughout. To identify a word of art is always a major accomplishment, for it indicates where a scientific concept has been isolated out

of familiar experience. In China, as in other cultures before the Scientific Revolution, most special terms were borrowed from ordinary language, not newly coined (as we now commonly coin them from obsolete words and from roots in dead languages). They are identifiable only because they were given new senses which they did not have in everyday speech. "*Tan* 丹," the word for cinnabar, was borrowed to refer to any alchemical elixir, even those which did not contain cinnabar. "*Fu* 伏," a common word meaning, among other things, "to subdue," was taken to refer to laboratory processes which rendered inorganic substances impervious to physical or chemical change, while in daily discourse the word had no application to minerals at all. On the other hand, if the idea of technical terminology is to have any meaning at all, one is not free to suggest a special sense when the everyday sense is fully applicable (unless, as is often the case, the technical sense is earlier). One would hesitate, for instance, to call "*shui* 水" (etymologically, "water") a special alchemical or medical term for "solution," since the word had the same meaning in common discourse.

A similarly ordinary word which recurs in the Mohist Canon in obviously technical usages is *cheng* 正, "straight." It appears to have two senses in two contexts (we have had to reject, at least provisionally, the attractive suggestion that in some cases it refers to the optical axis): (*a*) of objects and images, "upright" (B20–B22) in contrast with *i* 杝, "slanting," and *i* 易, "inverted"; (*b*) of mirrors, "plane" (B21) in contrast with *wa* (位→) 洼, "concave," (B22) and *t'uan* 團, "convex" (B23). This appears to be its sense in the highly technical phrase 過正, "go beyond the mirror plane" (B21, B23).

4. Finally, it is necessary to consider the feasibility of any proposed experimental arrangement in terms of (*a*) available apparatus, (*b*) experiences possible for an observer the size of a man, with human faculties, rather than for the theoretical dimensionless information-gathering device whose adventures we follow in so many university physics textbooks, and (*c*) reasonable development from simpler observations that could

have been made in mundane surroundings. It is the tendency to examine early scientific writings exclusively against a background of the correct modern understanding, rather than against the background of unevenly distributed ignorance out of which they emerged, which often causes modern investigators to pass up valuable clues to ancient scientific styles. Several heretofore unnoticed idiosyncracies of the Mohist optical theories will be pointed out later (pp. 123–124,138 and 147).

Our goal in pooling our efforts in this translation has been to apply all these standards for the first time. We have no illusions about having produced a definitive rendering; in fact several propositions have been interpreted so tentatively that we can be sure our versions will not stand the test of time. The sooner they are obsolete, the better we will consider our attempt to have been, for our point is to make as clear as possible not only what we know with confidence but where the centers of concentration of our ignorance lie.

The Optical Propositions

There are eight optical propositions, which we summarize as follows:[7]

B16. Connection of shadow formation with movement, implying a definition of shadow as the absence of light

B17. Shadows cast by multiple light sources, in which light falls upon a shadow without completely obliterating it

B18. Image inversion

B19. Use of a plane mirror to cast a shadow in the direction of the sun

B20. Variation of shadow size depending upon the size and distance of the source

B21. Perspective in mirror images(?)

B22. Image formation in a concave mirror

B23. Image formation in a convex mirror.

The Mohist optics is primarily the study of shadows. The

---

[7] For other propositions mistakenly supposed to be concerned with optics, see Appendix A to this chapter (p. 136) below.

word *ying* 景, 影, "shadow," appears in every one of the eight relevant Canons, the world *kuang* 光, "light," only in some of the Explanations. The Mohist, as one would expect of a pioneer of optics, knows nothing about the propagation of light except what he can learn from the silhouetted shapes and straight edges of shadows, and how they vary as the relations between object and light source change. In the present version we translate *ying* consistently by "shadow" and resist the temptation to shift to "image" when the Mohist turns his attention to mirrors. From his point of view nothing has changed except that, for reasons he does not or cannot explain, shadows on lustrous surfaces show up the detail as well as the outline of the object. He can explain the inversion of the image in a concave mirror only by what he has learned from the study of true shadows, that is, in terms of light obstructed by the object, so what he explains is the inversion of the silhouette (B22 and, on one hypothesis, B18). In spite of this he makes the very great achievement of explaining the inversion of the shadow by a geometrical figure which is not visible to the eye, a cone of shadow shrinking to the focal point and then opening out in a second cone which has rotated (*yun* 庫→ 運, B18) halfway round, so that the image is upside down. Shen Kua got no farther some 1300 years later, comparing the figure which he postulates to a drum narrow at the waist.

The Mohist speaks of the light following straight paths which intersect at the conjugate focus or at the center of curvature, two points which he does not distinguish (B22). Has he the concept of light rays? The only straight line in optics which is obvious to the eye is the edge of a shadow, cast for example by a wall in a strong light. The Mohist's intersecting lines and Shen Kua's waist drum, conceived rather than perceived, are the straight edges of cones of shadow. So long as he deals with shadows and silhouettes, the Mohist can offer not only observations but explanations, deriving their variations from the figure of the object which obstructs the light, relative dimensions, distance, inclination, reflected light, curvature of the reflecting surface, and single and multiple light sources, all depending on the principle laid down in the first optical Explanation, "Where the light reaches the shadow disappears." To an extent quite unusual in early Chinese science we find ourselves in a landscape of geometrically visualized figures, very far from the metaphysics of yin and yang and the Five Phases, of numerology, time cycles, and resonances. The same is true of the propositions on mechanics. Indeed the only reference in the Canon to yin and yang or Five Phases is a refutation of the theory that there are regular ascendancies among the latter (B43).

But does this tendency toward geometrization indicate a radical rejection of the conventional Chinese world view, whose phenomenal world was a self-regulating organism in which local changes are best explained by resonance between things categorically related? It should be clearly recognized that interpretation involves a great deal of unjustified extrapolation. Apart from the disputed first sentence of the Explanation of B18, there is no evidence that the Mohist recognizes a need for geometrical frames of reference except in the case of shadows. As for the rest, it is quite possible (and most reasonable in the light of what little is known about the process of scientific change) that he saw the world in terms of resonance and other conventional concepts of his time (this seems particularly likely in the very disputable B21). The Mohist innovations which can be clearly demonstrated, limited though they may be, were more than sufficient to guarantee incompatibility with the mainstream of Chinese natural philosophy.

Translation
We observe the following conventions in presenting the Chinese texts:
1. A character followed by an arrow and enclosed in parentheses appears in the current text, but we amend it to the character immediately following the parenthesis.
2. A character enclosed in square brackets [ ] appears in the current text, but we excise it.

3. A character enclosed in pointed brackets ⟨ ⟩ does not appear in the current text, but we insert it.

Reading the text while omitting everything enclosed in square brackets and parentheses gives our critical version, on which the translation is based. Textual notes are keyed to superscript numbers in Chinese text.

PROPOSITION B16

景不 (從→) 徙¹, 說在改爲².

景· 光至景亡· [若在盡古息]³

Canon: A shadow does not shift. Explained by: remaking.

Explanation: Where the light reaches, the shadow disappears.

TEXTUAL NOTES

"The shadow of a flying bird has never stirred 飛鳥之景未嘗動也" is found among the paradoxes of the sophists in *Chuang-tzu*, 33.77. It appears as a paradox of the sophist Kung-sun Lung in *Lieh-tzu* (ca A.D. 300), in a form directly modeled on this Canon and including the reference to the Explanation with which it ends: 景不移者, 說左改也 (*4*:7b2).

1. The confusion of these graphs is systematic, also found in A49 and B13. The correct reading survives in a quotation by Ssu-ma Piao (died A.D. 306), in a comment on the paradox as it appears in *Chuang-tzu* (SPTK, *10:* 4b5–7):

"The bird screening off the light is like a fish screening off the water. The fish moves and screens off the water, but the water does not move. When the bird moves the shadow is born; where the shadow is born the light disappears, but disappearing is not going away and being born is not coming. Mo-tzu says: 'A shadow does not shift 影不徙也.'"

2. For this phrase, see the Book of Odes (*Shih ching*), no. 75: 緇衣之宜兮, 敝予又改爲兮, "How befitting is the black robe. When it is worn out, I will again make a new one [for you]."

3. The interpretation of this phrase by Sun I-jang and others ("If it is present it remains forever") is unconvincing, since a compound *chin ku* 盡古, "forever," is otherwise unattested,

and it is not persuasive to give *hsi* 息, "cease," the sense of "remain." Illustrative phrases beginning with *jo* 若, "like, for example," recur throughout the Explanations and constantly present difficulties. We understand them to be originally a separate list of illustrations, keyed to the corresponding Explanations by references which were often misunderstood when the illustrations were subsequently incorporated into the Mohist Canon. When the phrase in question is understood in the light of the Mohist syntax and diction, it means roughly "For example, when being present in it is over, the past ceases." We suggest that it was meant to refer to the Explanation of B13: 南北在旦, 有在莫, 宇徙久, "Both north and south are present at dawn and again at sunset. The shifting of space involves duration." The illustrative phrase on this hypothesis carried the single word *hsi* 徙 to key it to B13; the scribe responsible for writing it in was looking at the Canons instead of the Explanations, and thus did not find a *hsi* until he got to B16: 景不徙. In all major misplacements of this kind except one, a similar misunderstanding of a reference may be identified.

PHYSICAL INTERPRETATION

A shadow does not actually move as its object (or light source?) moves, but the old shadow is obliterated and a new one formed. This is because as the light rays in relative motion impinge on the shadow they destroy it.

The Mohists were consistent defenders of common sense against the sophists, and it is at first sight surprising to find a well-known paradox of this school placed at the head of the optical sections. The reason is no doubt that the Mohist approves of the sophism as a vivid illustration of the idea that, although it is apparent that shadows move with their objects, they do not share the continuity of the object's motion (just as one knows from observation that they do not necessarily share its speed). The shadow, in other words, has the special qualities of an interference phenomenon, disappearing and reappearing as light enters or is impeded. It will be noticed that he does not give the argument for the paradox (of which Ssu-ma Piao's account is a very plausible reconstruction, although probably without ancient authority). He is not concerned with logic-chopping in this part of the Canon and no doubt assumes that the argument of the sophists is common knowledge. He merely lays down what for him is the important point involved in it, that "where the light reaches the shadow disappears." This would not be a truism, since in folklore the shadow has an independent and numinous existence. That the shadow results from obstruction of the light is essential, for example, to his explanations of shadow inversion in B18 and B22.

PROPOSITION B17

景二, 說在重.

景. 二光夾一光, 一光者景也.

Canon: Shadows are two. Explained by: redoubling.

Explanation: When two lights flank one light, whatever is illuminated by one light is in shadow.

TEXTUAL NOTES

The Mohist writers do not use the particle 者 casually, and there is a significant difference between the phrases translated "one light" and "whatever is illuminated by one light." Cf. A51: 一然者, 一不然者, "something so in one respect, something not so in one respect," or B65: 一法者, "Things to which one standard applies."

PHYSICAL INTERPRETATION

The key to understanding this proposition lies in the connection between the two shadows of the Canon and the "redoubling" exemplified in the Explanation.

A Canon in the scientific sections presents either a controversial thesis or a problem, so the first question to ask is why the Mohist should be surprised that an object can cast two shadows. The trivial observation that two shadows imply two lights is certainly not the point being made in the Explanation. The problem that concerns him evidently arises out of the conception of light destroying shadow developed in B16, according to which lights on opposite sides of an object might be expected to obliterate each other's shadows rather than both

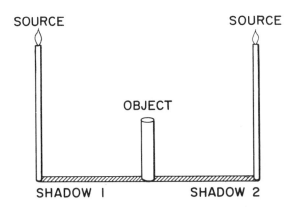

Fig. 6.1

casting shadows. His answer is that the spaces screened from one of the two lights show up as shadow against the areas illuminated by both. To prove this answer he takes the case of a light with lights on both sides, in which it is obvious that there is no total shadow and that the intensity of any visible shadow must be relative to "redoubled" light (Fig. 6.1). In the case of multiple light sources, there is a middle ground between the opposites light and shadow. This is no more alien to common sense than the idea that, although "open" and "shut" are mutually exclusive, a door may be partly open or partly shut.

According to another explication (originated by T'an Chieh-fu), this proposition was meant to account for the shadow and its penumbra cast by an extended light source, so that the doubling spoken of by the Canon is that of the shadow rather than of the source, but this view interprets the word 光, rendered above and elsewhere in these propositions as "light," to refer instead to the shadows. Such a use of this word is not completely unknown elsewhere in early literature, but in the *Mo-ching* the distinction is rigorous and fundamental. One may, with Hung Chen-huan, combine interesting features of both interpretations by not demanding that the sources be diametrically opposite, in which case there will be two partially illuminated shadows, with the area of overlap completely in shadow (see Fig. 6.2). The Explanation would then read somewhat less simply, and thus with less inherent likelihood, "When two lights flank [an object, part of the space between them will be] illuminated by one light. What is illuminated by a single light is the shadow [cast by one source only]."

Regardless of the details of the model, this proposition moves beyond the last by specifying the conditions in which light can fall upon a shadow without obliterating it.

PROPOSITION B18

景到¹在午²有端, 與景長. 說在端.

景. 光之 (人→) 入³煦⁴若射. 下者之 (人→) 入也
高, 高者之 (人→) 入也下. 足蔽下光, 故成景
於 (止→) 上⁵, 首蔽上光, 故景於下. 在遠近
有端, 與於光. 故景 (庫→) 軍⁶內也.

Canon: The turning over of the shadow is because the criss-cross has a point from which it is prolonged with the shadow. Explained by: "point."

Explanation: The light enters and shines [lit., "the light's entry is a shining"] like the shooting of an arrow (N.S.); *or:*

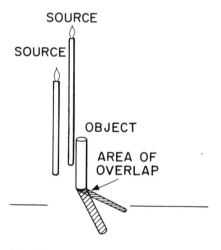

Fig. 6.2

The light's entry into the curve is like the shooting of arrows (A.C.G.).

The entry of that which comes from below is upward; the entry of that which comes from high up is downward. The legs cover the light which comes from below, and therefore form a shadow above. The head covers the light from above, and therefore forms a shadow below. This is because at a certain distance there is a point which coincides with the light. Therefore the revolution of the shadow is on the inside.

TEXTUAL NOTES

1. Equivalent in meaning to 倒 (Morohashi 1950 sense 3), again in B21.

2. Morohashi 2703 sense 4, "criss-cross 一從一橫."

3. Discussed later.

4. *Hsu* 煦 (also A47, written 昫) is here taken by Sun in the sense of "emanate," although it may be objected that the word suggests an emission of warmth rather than light. Most prefer to emend to 照, "shine," on the basis of close graphic similarity (T'an, Kao, Wu, followed by N.S.). But it is worthy of attention that *hsu* was a Sung taboo (as the personal name of Che-tsung 哲宗, r. 1086–1100), and that Sung taboos from the Taoist Patrology of 1111–1117 are only partially restored in the Ming text (Ch'en Kuo-fu, p. 189); Luan T'iao-fu notes two in *Mo-tzu* (p. 155). Although Ch'en Yuan in his monograph on taboos (p. 641) does not mention a regular substitute character for *hsu*, the one used by Shen Kua (item 29) was 旬. This suggests the alternate hypothesis (A.C.G.) that the original word both here and at A47 was *kou* 句 (= 勾, "curve," Morohashi 3234 sense 3.1), later mistaken for the substitute character (on this kind of over-correction, see Ch'en Yuan, p. 607). In view of the tendency in the Mohist Canon for graphs to collect unusual or corrupted radicals, the chance that the word is *kou* is strong in any case.

5. Emended on grounds of parallelism.

6. Sun and others emend the graph to 庫＝障, "to block," both here and in A48. But Wu Yü-chiang, although he accepts the unemended graph here (as *k'u*, "storehouse," used to refer to a

dark room), shows convincingly that in A48 the word is *yun* 運, "rotate," written without its radical; on bronze inscriptions the graphs 軍 and 庫 both appear as 厙 and also as 軰 (Tuan Wei-yi, pp. 247, 649).

This section presents syntactic and textual problems which require extended discussion. The syntactic difficulties center on two words:

(1) *Tsai* 在 (two occurrences). This cannot be the preposition ("in") common from the Han onwards. Elsewhere in the Mohist Canon it is the full verb, "is in." But here in both cases it seems to cover the whole succeeding phrase, in the cognate sense "depends on, is because of." That the succeeding phrase would be verbal and not nominal is confirmed by the following example (*Mo-tzu* 8.3): 是在王公大人爲政於國家者不能以尙賢事能爲政也, "This is because kings, dukes, and great men who exercise government in a state cannot exercise government by promotion of worth and employment of ability."

(2). *Jen* 人 (three times). In all cases it seems to occupy a verbal position; it is impossible to take *chih* 之 as the verb ("go to") since this usage is not found in the Mohist Canon, which is very consistent in point of diction. Thus interpretations of the sort "When the light reaches the man . . . " may be ruled out. The characters 人, 入 are so often confused throughout the *Mo-tzu* that at some stage, apparently, they were indistinguishable except by context.

Fortunately all disputable words appear elsewhere in the Mohist Canon. *Tuan* 端, the starting point or end point of a measured length, is defined in A61. *Tao* 到 in the sense of 倒, "turn over," appears in B21; it refers to the event of turning over, not the state of being changed round (*i* 易, A48 and B22). In other optical sections we find 長, "prolong" (B22, of the prolongation of the rays after intersection); 遠近, "far, near" (B22 and B23, of distance along the axis of a curved mirror); 內, "inside" (B22, of the space between the center of curvature and the mirror). The two most problematic graphs (煦=昫, 庫) occur, together with the *i*, "changed round," of B22, in two adjacent sections in the series on change and motion (A47, A48). These very difficult sections are discussed in Appendix A below. We conclude that there is little doubt about Wu's recognition of the last of the graphs as representing *yun*, "rotate." Here the Mohist is conceiving the shadow in the form of two cones with touching apexes (the figure which Shen Kua later compared to a waist drum), with the inner cone rotating through 180°. The evidence of A47 in the case of *hsu* is less firm but, such as it is, favors the theory that the word in question is *kou*, "curve."

PHYSICAL INTERPRETATION

Corruption and syntactic obscurity account for the very divergent explanations which have been offered for this proposition. Having applied the principles upon which our approach is based, we find that there remain two possible interpretations (corresponding to the readings discussed in note 4 immediately preceding) which would be consonant with the physical resources of the Mohists. Although one of us (N.S.) favors the first and one (A.C.G.) the second, we are unable to extend the process of elimination rigorously to either, and it must be emphasized that except on impressionistic grounds they must be considered equally likely until some student of the Canons produces tangible new evidence.

On the first reading, hitherto generally accepted among those concerned with the Mohist contribution to the growth of physics, this proposition refers to the camera obscura with pinhole aperture. Wu Yü-chiang even finds technical terms for the darkened room (庫) and the screen (長→帳), even though there is no evidence that the Mohists devised special experimental equipment; their balances, pulleys, ladders, and mirrors of various kinds were ready at hand. Wu did not, however, find a technical term for the pinhole, and without it we can only be less than certain.

Every modern commentator attracted to the hypothesis of the camera obscura has complicated it unnecessarily by the general assumption that the Proposition is concerned with a reflected-light image—apparently because this is the type of image generally associated with a pinhole camera in modern

Physics textbooks. In experimental situations, however, such images are immensely harder to observe, and thus to discover, than silhouetted images. These would also be "shadows," the primary meaning of the word 景, which in Propositions B22–B24 is applied to mirror images, or at least their outlines. Just as there is every practical physical reason to suspect that here the reference is to an image between the light source and the pinhole, it makes sense of the text. Otherwise (1) the sentences "The legs cover the light . . . forms a shadow below" must be laboriously interpreted (for example, by Hung Chen-huan) as a mere statement that the image is inverted; and (2) the unambiguous assertion that the head and legs *cover* the rays has to be ignored, and (3) the word translated "revolution," and understood by us to describe the action by which the image enters the pinhole and simultaneously reverses and inverts itself, has to be emended to 庠 and then given the very unlikely sense "is bright"; the last clause thus becomes both syntactically suspect (no preposition before "the interior") and logically a *non sequitur* ("is bright" would thus be meant only to assert "exists"). Although a reflected-light image cast by a pinhole can be seen only by an observer inside a very well-darkened room (the camera obscura), the ability of a pinhole to form an image of an object silhouetted against the sun or another bright light could have been discovered at least as easily with the aid of nothing more than a sheet of thin opaque material pierced by a pinhole, and a piece of translucent material (say thin silk) to serve as a screen. If, on the other hand, we would do better to detach the character 勾 from the accretions of radicals which disguise it, we have no choice but to understand that the Mohist is explaining the principle of image inversion in the concave mirror, a principle which he applies in B22. The Explanation, except for the pivotal first sentence, can be translated precisely the same as in the other case. His comparison with the shooting of arrows may refer to more than the straightness of the light's motion. It may imply that all the light enters the curve of the mirror along the axis, as an arrow passes along an axis perpendicular to a bow. His comparison with "shooting 射" is perhaps with the character itself, originally a picture of a bow and arrow 弓. If this is true, it would follow that all paths intersect at the center of curvature 中 (a concept which he does not distinguish from that of the conjugate focus, as will be seen in B22). It would also follow that the shadows cast by head and feet screening off the light which silhouettes them are reversed before they reach the mirror (even a small concave mirror will give a full-length image).

Although the demands of communication with modern readers dictate the use of the word "rays" in this interpretation, it is well to be aware that the text of the *Mo-ching* is not so explicit; its explications are in terms of "the light." This issue will be considered further in our commentary to Proposition B22.

PROPOSITION B19

景迎 (日→) 日.¹ 說在 (博→) 轉.²

景. 日之光反燭人, 則景在日與人之間.

Canon: The shadow cast in the direction of the sun. Explained by: reversion.

Explanation: When the sun's light, reflected, illuminates a man, his shadow will be located between the man and the sun.

TEXTUAL NOTES
1. There is the same confusion of graphs in B10 and B47.
2. The Mao K'un 毛坤 edition of 1581 reads 博, of which the Taoist Patrology reading here reproduced is certainly a corruption. We find the phonetic 專 corrupted to 尃 elsewhere in the Taoist Patrology version (A81, B57, B62). In all cases except the first, the correct version is preserved in the Mao edition. Sun I-jang is no doubt right in identifying the character as 轉, found twice elsewhere in the Mohist Canon (A94), written once with its usual radical and once as 傳.

For the phrasing of the Canon, cf. *Kuan-tzu*, ch. 52 (*16*: 9bl): 意者君乘駁馬而洯桓, 迎日而馳乎, "Were you perhaps riding round on a piebald horse and galloping in the direction of the sun?"

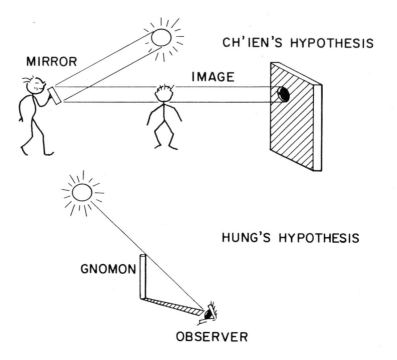

CH'IEN'S HYPOTHESIS

MIRROR

IMAGE

HUNG'S HYPOTHESIS

GNOMON

OBSERVER

Fig. 6.3

PHYSICAL INTERPRETATION

Although it is difficult to be completely sure that no other experimental situation is meant, this proposition probably refers to the use of a small plane mirror to reflect the sun's light onto a man and thus cast his shadow onto a surface not illuminated by full sunlight (Fig. 6.3, top).

The word translated "cast in the direction of" above can carry the sense of either "facing" or "going to meet someone or something which is itself approaching"—in this case the sunlight. Because Ch'ien Lin-chao, whose overview of this proposition we follow, took it in the former sense (despite the fact that the Explanation was clearly interpreting it in the latter sense), he left himself open to the objection that for the shadow to be characterized as "facing the sun" it would have to be substantially full length, impossible with the small mirrors of the time. On the strength of this objection Hung

Chen-huan chose a different emendation ("摶" to "捶"), so that "reversion" becomes "sighting along an upright," and took the proposition to be concerned with an eye placed at ground level at the tip of a gnomon's shadow, so that it sees the sun lined up with the end of the shaft (Fig. 6.3, bottom). Although a gnomon was used at the time to sight transits of the moon and stars across the meridian, Hung is unable to interpret the Canon as a declarative proposition. He also has recourse to a very unlikely reading of one character in order to get "When the sun's light is unable to illuminate a man . . . ," whereas if his understanding of the situation were correct it is only the pupil of the observer's eye, not his whole body (as this reading necessarily implies), which is covered by the gnomon's shadow.

PROPOSITION B20

景之小大，說在 (地→) 杝[1]缶[2]遠近·

景· 木杝景短大，木正景長小· (大→) 火[3]小於木，
則景大於木· 非獨小也，遠近，[4]

Canon: The size of the shadow. Explained by: tilt and distance.

Explanation: When the post slants the shadow is shorter and bigger; when the post is upright the shadow is longer and smaller. If the flame is smaller than the post the shadow is bigger than the post. It is not only because it is smaller, but also because of the distance.

TEXTUAL NOTES

1. Emended from the Explanation. The latter graph is used regularly in the Canon for *i* 迆, "slant" (Morohashi 14478 sense 2, which however notes the interchange only in the sense of "split"). See B21 below for a second example.
2. A graphic variant of 正 common in the dialectical and military chapters of *Mo-tzu* (Sun I-jang, p. 196.1). Further examples in B22 and B23.
3. In emending 大 to 火 we follow T'an Chieh-fu and others. Sun I-jang's 光, "the light," is hardly less inherently likely.
4. Even though it is possible to make sense of the last sentence in context, it gives an impression of being mutilated.

PHYSICAL INTERPRETATION

We take the final sentence of the Explanation to mean that shadow size depends on both size and distance of the source. The understanding of the rest of the Explanation hinges on the meaning of "bigger" and "smaller." The traditional interpretation (due to Sun I-jang) takes them as "clearer" and "less distinct," with reference to the definition of the shadow. A more likely explication (by Hung Chen-huan) takes them as "wider" and "narrower," referring to the width of the shadow of the gnomon's top. The Canons do not use the common words 廣 and 狹 "wide" and "narrow," although they are found in the military chapters. The character 廣 is used in the Mohist Canon only for breadth as opposed to length 脩 (B4). There is consequently little doubt that the width of the shadow would be described as 大, which seems to be used generally when the dimension is not length, for example, in A55: 厚, 有所大也, " 'Thick' is having size."

Both Sun and Hung make the mistake of not checking their assumption that the text refers to the shadow of a gnomon (like that used in early astronomy) cast on the ground. But Hung's explanation would actually be correct only for the case where the shadow is cast on a surface perpendicular to the ground, and even then only when the light source is higher than the upright gnomon, and when the angle of tilt is small (Fig. 6.4). Hung's assertion that the shadow of the tilted gnomon cast upon the ground is shorter and wider than that of the upright gnomon is erroneous; the shadow width is, with a point source, always maximal when the gnomon is perpendicular to the ground. With a common source of sensible width such as a candle flame, the apparent width of the shadow's tip upon the ground does not vary appreciably as the gnomon is tilted. Nor does the definition of the shadow (upon the ground or upon a perpendicular surface) vary appreciably when the gnomon is tilted except that, if the tip of the gnomon approaches the flame very closely, the increase in relative size of the source will widen the penumbral area at the expense of the central shadow.

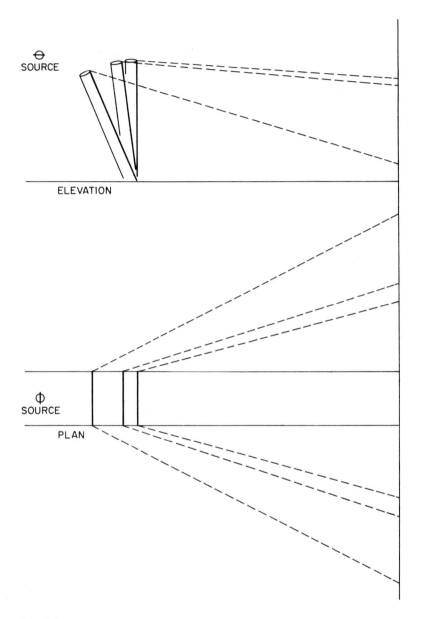

Fig. 6.4

The Explanation clearly refers to the case in which the gnomon is tilted toward the light source, but in principle the Canon could be referring to (1) a tilt in either direction, with the shadow thrown upon the ground, or (2) the same experimental arrangement as the Explanation, but with the light source below the gnomon.

PROPOSITION B21

臨鑑而立, 景到, 多而若少. 說在寡區.

臨. 正鑒景寡, 貌能[1]白黑遠近杝正異
於光. 鑒景當, 俱就去[2]. (亦→) 不當,
俱 [俱] (用→) 向北[3]. 鑒者之臭[4]於鑒
無所不鑒, 景之臭無數而必過正,
故同處. 其體俱然, 鑒分.

Canon: If one stands upright looking down at the mirror the shadow turns over, and the more there is of it the less there seems to be. Explained by: reduced area.

Explanation: In a plane mirror the shadow is reduced. The appearance, shading, distance, and inclination differ from those of [the object in] the light.

When mirror and shadow are aligned, [object and shadow] approach and withdraw together. When they are not aligned, [object and shadow] turn toward or away from each other.

The *ch'ou* of the man looking at himself are mirrored in the mirror without exception. The *ch'ou* of the shadow are numberless but must go beyond the mirror plane; they are therefore [squeezed into] the same places. Since this is so of all the parts of his body, it mirrors their portions.

TEXTUAL NOTES

1. 能＝態 (Morohashi 29454 sense 10), a graph not found in *Mo-tzu*, although we do once find 能 used for it (52.15).

2. The punctuation of this and the next sentence presents difficulties. *Chü* 俱 in the Mohist Canon is generally adverbial, meaning "all, both" (referring backward to the explicit or implicit subject), so that the last three words are most naturally taken as a phrase: "they approach and recede together." If

so, the first three words are also a phrase, with *tang* 當, "correspond," as an intransitive verb with two subjects. Cf. *Kuan-tzu*, ch. 55 (*18:* 3b4): 名實當則治, 不當則亂, "If name and object correspond it will be orderly; if they do not correspond it will be unruly."

3. This mutilated sentence looks as though it should be parallel with the preceding one. The graph 介 has a variant 企; neither is found elsewhere in *Mo-tzu*. We emend conjecturally to 不, on the grounds that if the same verb (*tang*) recurs in parallel sentences it is natural to guess that it will be negated, as in the *Kuan-tzu* passage just above. The impossible repetition of the adverb *chü* requires the excision of one. The last two words should form a pair of opposites parallel with "approach and recede"; we therefore follow Kao Heng in reading 向北 (=背), "face towards and turn one's back on," a pair of opposites of which several graphic forms are recorded (Chu Ch'i-feng, p. 1911).

4. Finally, the crucial character 臬 has never been convincingly amended. T'an's correction to 臬, "target peg," is an attractive guess (cf. the comparison of light with a flying arrow in B18). Other suggestions are 臭 (Sun I-jang), 夏 (Kao Heng), 兒 (Wu Yü-chiang), and 異 (Chang Ch'i-huang); the most recent is 臬 (Li Yü-shu). None of these is strongly supported, nor can we be at all sure from the context what the character, whichever it may be, must mean. In order to leave open the question of what character should occupy this space, and to make explicit the paucity of clues offered by the context regardless of how the Explanation as a whole is interpreted, we find it advisable to leave the Chinese word untranslated.

There is, however, some evidence which we may offer in favor of Li Yü-shu's conjecture. In *Shih chi* (*23:* 9.1) there is a case of *ch'ou* 臭 as a mistake for *tse* 澤, which must have been written with its old graph 臬 and mistaken for the more common graph with two strokes obliterated (Chu Ch'i-feng 1187). *Tse* is perhaps the most common early word for the luster of a reflecting surface. Its usage may be illustrated from certain Confucian comparisons of various virtues with qualities of jade as in *Hsun-tzu* (30.8): 溫潤而澤, 仁也, "Being lustrous with a warm glossiness is benevolence"; *Chia-tzu hsin shu* 賈子新書 (*B:* 37b1): 澤者鑑也, 謂之道, "In that a lustrous thing serves as a mirror, one calls it the Way." In the present case the adjective would be in a nominal position, "what is lustrous":

Everything lustrous in the man looking at himself is mirrored in the mirror without exception. The lustrous features on the shadow are numberless, but must go beyond the mirror plane; they are therefore (squeezed into) the same places. Since this is so of all members of the body, it mirrors their portions.

This accords with one's impression, strengthened by the next two sections, that for the Mohist the reflected features of the face are additional to the *ying*, which even in the mirror is still the shadow, the outline of the image.

5. The phrase reappears in B23. See the late attested phrase 矯枉過正, "bend the crooked beyond the straight, over-straighten" (Morohashi 24015..78).

PHYSICAL INTERPRETATION

Our interpretation is highly tentative, for this is the most problematic and ravaged of the optical propositions, and the one most generally given short shrift by commentators. The most popular tradition of expounding the text, in terms of convex or concave mirrors, is seriously deficient in point of both philology and optics. A major advance in thinking out a feasible physical situation was the hypothesis of T'an Chieh-fu that the text has to do with two plane mirrors joined at an angle, but his explication of the text was marred by several extremely forced emendations (as will appear anon). It did not, in any case, fit the Canon, although there is every reason to believe that the writers of the Explanations understood the Canons. Hung Chen-huan, attempting to apply the hypothesis of two mirrors face to face, with the object between, inadvertently left no role for the observer, and even so relaxed the high standard of skepticism which usually led him to demand consistent and soundly precedented interpretations of characters. For instance, 臬 is taken first to mean "placing face to face" on the basis of the classical sense "aim at a target," and then later to mean "a peg." Hung was quite unable to offer his usual word-for-word paraphrase of the last part.

Since this proposition is the first of a series which deal with the formation of images in mirrors (as we have pointed out, B18 has to do with the special question of image inversion), it is most reasonable that it deal with plane mirrors. Neither T'an nor Hung was able to account for the fact that the text does not indicate in any way that more than one mirror was meant. T'an took 正 in one occurrence as synechdoche for "two plane mirrors 二正鑑," a measure to which such a good philologist could have been driven only by sheerest exasperation.

But T'an's recognition of 正鑑 as a technical term for a plane mirror was correct; it has been followed in such important non-Chinese studies as those of Needham (IV. 1, 83) and Yabuuti (p. 445, relying mainly on Kao Heng). We diverge from T'an at the point where, since the text does not specify multiple mirrors, we refrain from positing more than one; this point had been reached by both of us long before we compared our interpretations. Of the many experimental situations we have evaluated, only one (first proposed by A.C.G.) has survived elimination on philological grounds. The chief philological constants, in addition to those already established in our textual notes, are these: 當 in physical contexts in pre-Han literature ordinarily means "face to face with, in alignment" (相當, "opposite"; 當路, "plumb in the middle of the road"), not "be placed at an angle" or "meet." We were thus forced to reject an interpretation in which, by moving a plane mirror, the object and its reflection can be made contiguous. Although we find T'an's suggestion that *cheng* 正 is a technical term for the optical axis most attractive, after attempting to apply it consistently to all occurrences in B21, B22, and B23 we find that it is not needed, and leads to interpretations which do not stand up for other reasons. We take *cheng* (used of object or shadow) in the sense of "upright" or (used of mirrors) to mean the mirror plane, or the tangent plane of a concave or convex mirror. These last senses are not so recondite as they seem, since the mirror plane or tangent plane is perpendicular to the ground when the mirror is held up. Thus the equivalence of 正鑑 and "plane mirror" is quite literal.

We propose, in sum, that this proposition is attempting to formulate laws of, and an explanation for, perspective. The Canon is talking about the image one sees of oneself when standing over a mirror on the ground, an image both inverted and receding. The remainder we take as an attempt to explain the fact that one's head appears to be much smaller than one's feet, since the principle of inversion has been dealt with already in B18. We read the Explanation as concerned throughout with the same problem. First it notes the fact that object and image do not coincide in size (nor in other variables). Secondly it notes the connection of recession with motion along the optical axis, although without making the latter concept explicit. The letter of this part of the Explanation is best understood as referring to what happens when, beginning with a mirror squarely in front of the observer, he gradually tilts the mirror; the image tilts away from him, and begins to recede. The last part, we can be rather sure, develops the explanation given in the Canon. We are less sure just what explanation the words are meant to convey, but they appear to be saying that as the mirror is tilted the image is compressed into a smaller area, so that separate positions on, or parts of, the object must occupy the same position in the image, and thus no longer be separately visible.

Then what are the *ch'ou?* They could be some sort of spatial minima, or simply the discrete parts of the object (eyes, nose, mouth). Whether either is correct we must, like many other questions raised by this proposition, leave in suspense. But perhaps we can simplify further studies by stating clearly that we do not find any suggestion of atomism in B21. An atom is, first of all, physical rather than spatial. More important, it must represent the smallest physically conceivable division of matter; an arbitrarily tiny unit thought of as further divisible in principle is not an atom in the classical Democritean sense, and in fact fits philosophies which find matter continuous rather than atomistic in nature.

PROPOSITION B22

鑑 (位→) 洼[1], (量→) 景[2] 一小而易, 一大而
缶. 說在中之外內.

鑒. 中之內. 鑒者近中則所鑒大, 景亦大.
遠中則所鑒小, 景亦小, 而必正. 起
於中, 緣正而長其直也.
中之外. 鑒者近中則所鑒大, 景亦大.
遠中則所鑒小, 景亦小,而必易. 合
於〈中...〉[3]而長其直也.

Canon: When the mirror is concave the shadow is at one time smaller and inverted, at another time larger and upright. Explained by: outside or inside the center.

Explanation: *Inside the center:* If the man looking at himself is near the center, everything mirrored is larger and the shadow correspondingly larger. If he is far from the center everything mirrored is smaller and the shadow correspondingly smaller. But it is sure to be upright. This is because [the light] opens out from the center, skirts the upright object and prolongs its straight course. *Outside the center:* If the man looking at himself is near the center, everything mirrored is larger and the shadow correspondingly larger. If he is far from the center everything mirrored is smaller and the shadow correspondingly smaller. But it is sure to be inverted. This is because [the light] converges at the center... and prolongs its straight course.

TEXTUAL NOTES

1. 位 is probably a graphic error for 洼, the word used (in the form 窪) for the concave mirror by Shen Kua more than a millennium later (see Appendix B; he also uses the word "正" to refer to the upright image). This emendation (Chang Ch'un-i and others) is more plausible than others which have been suggested: 低 (T'an), 弧 (Kao Heng), or 區 (Chang Ch'i-huang).

2. Parallels between B22 and B23, which deal respectively with concave and convex mirrors, indicate that 量 in this Proposition should be corrected to 景.

3. Parallels between the second and third paragraphs of B22 justify restoration of 中 in the otherwise incomprehensible last

sentence of paragraph 3, and suggest that two characters are missing from the same place.

PHYSICAL INTERPRETATION

The recognition that this proposition refers to a concave spherical mirror came early in the history of its investigation, for it states the principle of magnification clearly and correctly. As Ch'ien Lin-chao has explained, an observer moving toward a concave mirror would see an increasingly larger upright image of himself until his eye reached the center of curvature. Then he would see nothing until his eye had advanced beyond the conjugate focus, for the image at intermediate points would be aerial. From the conjugate focus inward, the image would also be as described. "Center 中" is thus a technical term which includes both center of curvature and conjugate focus—whose functions were not distinguished—and necessarily the portion of the axis which lies between.

What is the difference between *chien che* and *so chien*, literally "the one that mirrors" and "what is mirrored"? Both phrases reappear in B23; we may also compare B21 鑒者之臭於鑒無所不鑒, literally "The *ch'ou* of the one who mirrors are mirrored in the mirror without exception." *Chien che* means (or denotes) the man using the mirror, but in each case the context requires that it apply to (or connote) the object mirrored. This is not surprising if we take the simple situation in which a man looks at his own reflection, and when he is described as near or far from the center we can visualize him as moving his face toward and away from it (or his finger, as in Shen Kua's later account). We therefore render *chien che* by "the man looking at himself." The problem of *so chien*, "what is mirrored," is more difficult. If it is not the object mirrored what can it be? Not the shadow, from which it is firmly distinguished both here and in B23. The Mohist's answer, we propose, is that the shadow (*ying*) in a mirror differs from that on the ground or on a wall in that inside the silhouette eyes, nose, mouth are mirrored. Why this should be so is a question which, as we have noticed, is outside the frame of reference of the Mohist optics. The distinction is also implied in B21, which on the present interpretation notes how the separate features are squeezed into each other by foreshortening as the shadow tilts backward. In B22 and B23 the point would be that as the object advances or withdraws along a line perpendicular to the mirror the mirrored parts enlarge or diminish conformably with the outline of the whole. We therefore translate *so chien* by "everything mirrored," and understand it to refer to all the various parts of the observer's image (eyes, nose, fingers) as contrasted with the continuous line of the silhouette.

We have given serious consideration to Hung Chen-huan's identification of *so chien* as a technical term meaning "magnification." The idea that the Mohists had worked out this concept is most attractive, and it satisfies the physical necessities of its various contexts in B22 and B23. We have finally reluctantly rejected it as unsupported, since we can evolve no solid line of reasoning to lead from the meaning "what is mirrored" to "magnification," can find no examples of analogous transitions of sense in the formation of other technical terms, and can find no other indications of this concept in traditional Chinese optics.

The only optically problematic portion of this proposition is the two explanations of image formation, since one hesitates to assume gratuitously that the Mohists were thinking about optical phenomena in terms of a more or less geometrical analysis of ray propagation (see Fig. 6.5). But this hypothesis, advanced by Fang K'ao-po, is the only one which does not take serious liberties with the realities of syntax and diction. We accept it with the understanding that the Mohist is applying it only to rays that define borders between light and shadow. Because Fang interpreted "長" as "prolongation" rather than "extension," Hung rejected his whole interpretation (since when the observer is outside the center the rays are not, as Fang claimed, prolonged back of the mirror), but this was too abrupt. Hung's own explication, which takes "begin 起" and "meet 合" as technical terms used to distinguish distances as measured from the front and back of the mirror respectively, was hardly compelling, all the more because he had to set aside the sense

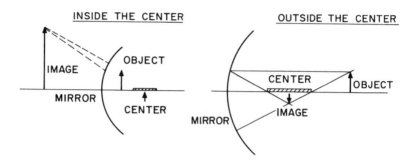

Fig. 6.5

of "center 中" which he used earlier in this proposition, interpreting it here as "the intersection of the optical axis with the
mirror's surface."

Finally, it is easy to overlook the aspect of the Explanation
in which it differs most fundamentally from modern optics: it
assumes that the light converges toward the center before reaching the mirror, not after being reflected back from it. In the
first case, where the object is between the center and the mirror,
the light is already "opening out from the center" as it passes
the object. In the second, where the object is outside the
center, the light first converges at the center, where what we
would call the rays from the object turn over (cf. B18, on the
hypothesis that it refers to the concave mirror), and then move
on to form the image. The Mohist's is of course the common-
sense viewpoint. It would require a considerably more advanced
optics to recognize that when reflected back the image is not
yet inverted, in spite of the fact that we *see* it as both inverted
and on or near the mirror surface. On his erroneous assumption
the Mohist is able to explain both the size of the shadow and
whether or not it will be inverted. In two sentences marked as
explanatory interruptions in the description by the final 也
("It's that...") he calls attention to the convergence and
divergence of the light and the straightness of its course. These
two facts would explain all the variations he has noted: (1)
Whether or not the image is inverted will depend on whether
or not the light has already converged and diverged when it

reaches the object; (2) the closer the object is to the center of curvature the wider will be the angle of the light; distance from the center will therefore explain the size of the shadow, provided that the lines are straight.

It is natural that an early investigator should take it for granted that the light contracts to a focal point and then opens out again before reaching the mirror; but has the Mohist any explanation of why this happens? Shen Kua, who seems to share the same assumption more than a thousand years later, explains the inversion on the analogy of an oar, the two ends of which move in opposite directions because in between it is trapped by the rowlock; he conceives an invisible surrounding obstruction through which the shadow has to squeeze at the focal point, as through the pinhole of the camera obscura (Appendix B). Whether the Mohist has such an answer depends on how one understands the disputed B18. If we take it to refer to the concave mirror rather than the camera obscura the first sentence, which on this interpretation reads "The light's entry into the curve is like the shooting of arrows," very tentatively suggests an answer. The comparison may apply only to the straightness of the light's course, as has been commonly assumed. But it might also mean that all paths by which light reaches a concave mirror are along the axes of the curve (like the arrow across the drawn bow), and therefore intersect at the center of curvature. This would fully account for the phenomena as the Mohist sees them. It would also imply that he has taken the first step toward a ray theory. Anyone who observes that shadows have straight edges and appear where light is interrupted is in a position to realize that light travels straight. We are not ready to credit the Mohist with geometrizing paths of light which are not the borders of shadows.

PROPOSITION B23

鑑團, 景一 ＜小＞/(天→) 一大[1]而必缶. 說在得.[2]

鑒. 鑒者近則所鑒大, 景亦大. (亦→) 其[3]遠, 所鑒小, 景亦小, 而必正. 景過正, 故招.[4]

Canon: When the mirror is convex, the shadow is at one time smaller and at another time larger, but it is sure to be upright. Explained by:...

Explanation: If the man looking at himself is near, everything mirrored is larger and the shadow correspondingly large. If he is far, everything mirrored is smaller and the shadow correspondingly small. But it is sure to be upright. The shadow goes beyond the plane and therefore recedes at the edges.

TEXTUAL NOTES

1. The Canon was broken, at the point marked by the slant line, when B14–B20 and B21 through the first half of B23 were transposed. It was long assumed that the second half is a separate Canon; the discovery that the parts can be spliced by the emendation of 一/天 to 一小一大 is one of Luan T'iao-fu's services to the study of the Mohist Canon (p. 87). As Kao Heng notices, the characters 一大, written one above the other, have run together to make 天. Comparison with the parallel Canon of B22, and of each Canon with its Explanation, confirms the emendation.

2. The final word in the Canon is evidently corrupt, and there has been no estimable attempt to explain it as it stands (see Kao Heng, pp. 138–139). We have no evidence for its restoration.

3. *Ch'i* 其, presumably written with the archaic graph 亓, is corrupted to 亦 some eighty times in the *Mo-tzu*. See Sun I-jang, p. 194.2.

4. 招 *T̂IOG/*chao* has been variously emended to 橋 (Sun), 杝 (Liang), 倒 (Chang Ch'un-i, Wu), 柖 (T'an), and 攝 (Liu). In *Mo-tzu*, 招 appears only for 喬 *G'IOG/*ch'iao*. See Sun I-jang, p. 3.3: 甘井近竭, 招木近伐, "The sweetest well is nearest to exhaustion; the loftiest tree is nearest to being chopped down." The only other examples are p. 24.3, where it is equivalent to 韶, the ancient Shao music, and Proposition B3, where the Explanation reads 抬.

But *ch'iao* does not precisely mean "lofty." The *Shuo wen* defines it as "high and bent 喬, 高而曲也," and the *Erh ya* 爾雅 as "branches bending upward," in contrast to *chiu* 杻, "branches bending downward." The two senses of *ch'iao* are unambiguously tied together in the *Shih ming* 釋名, where the word is defined as "a high mountain tapering to a peak, like the form of a lofty tree 橋." (For these definitions, see Morohashi 3990 and 14436). Of course one can play no end of elegant games with definitions dredged out of archaic lexica, and we would be much more convinced if we could find clear examples in actual texts. But 招 does seem a most appropriate word to describe the distortion of the image in a convex mirror, the edges of the image dwindling or receding like the branches of a poplar seen from below.

PHYSICAL INTERPRETATION

Here again the fundamentals of image construction are plainly set out, this time for a convex spherical mirror. The recession of the image at the edges is apparently being related to the increasing distance of the edges of the mirror from the tangent plane 正.

Appendix A. Other Propositions Which Have Been Said to Concern Optics

In his study of the Mohist optics Needham translates not only B16–B23 but also two additional Propositions (IV.1, 81–86): A48—which with A47 contains no less than three key terms of the optical propositions we have examined—and B56. Both are extremely obscure, but not utterly impenetrable. Although their relevance is plain, it does not follow that A48 any more than A47 is itself concerned with optics. The two Propositions stand, in fact, in the middle of a series of definitions, between "begin," "transform," "reduce," (A44–A46) and "move," "stop" (A49, A50)—terms germane to physics but not specific to optics.

PROPOSITION A48

(庫→) 運，易也.

庫. 區穴 [若斯] 貌常.

Canon: To "rotate" is to change round.

Explanation: The circumscribed space is constant in appearance.

TEXTUAL NOTES

We have already noticed (B18) Wu Yü-chiang's evidence for identifying the word which is the subject of A48 as *yun,* "rotate." The puzzling phrase *ch'ü hsueh,* "circumscribed space (?)," is also found in A63, where the context at least establishes it as a geometrical term. The bracketed phrase in the Explanation cannot be understood as "like this," since *ssu,* "this" does not belong to the language of any part of the *Mo-tzu.* The phrase seems to be another misplaced illustration of the type noted in B16. We suggest that it belongs to the definition of "reducing" in A45, where it would mean "for example, cutting off."

Definitions of this vague kind are generally contrastive, belonging to series of differentiated terms (for example the four terms connected with knowing in A3–A6). In rotation, of which the characteristic example is the turning potter's wheel 運鈞 (*Mo-tzu,* 35.6, 37.1) the parts replace each other inside a circumscribed space which remains constant. "Rotation" is therefore equated with *i* ("to change round"), the word applied in B22 to the inversion of the shadow. Another distinct kind of revolution is defined in the immediately preceding A47. This is *hsuan* 還, 旋, "to circle," written there with an unusual graph (Morohashi 1188 sense 5).

PROPOSITION A47

儇, 稹柢〈也〉.

儇. 昫民也.

In the Canon the first graph of the definition is unknown to Morohashi and the second unknown in any conceivably relevant sense. Sun I-jang noticed the odd fact that they resemble the first two graphs of the Explanation, the first phonetically and the second graphically. He suggested that both are corruptions of one sentence, 俱柢, which he interpreted as "anywhere is the base," that is, anywhere on a circle is its starting point. But the repeated radical in the Canon suggests a binome, and its reproduction in the Explanation with different graphs is hardly comprehensible except as a clarification of an unfamiliar binome by rewriting it in a more familiar form.

It may be noticed that Radical 115 has replaced the obsolete *Shuo wen chieh tzu* Radical 220 禾, said to represent a bent treetop incapable of straightening itself. Three of the five graphs with this radical are scarcely known except in binomes so written in the *Shuo wen* definitions themselves 稽枏, 稽枏. The first of these is found elsewhere written with a variety of graphs read *chih chü*/*ȚIEG-KIU or *KIĚG-KIU 枳枸, 枳句, 枳棋, 枝拘, as the name of a tree with curling branches or as a verb "to curve round, to swerve," used, for example, of the flight of a swallow (Chu 1282). It would seem that the binome of A47 means "to move in a curve" and is one of several in which the significant syllable is generally written with graphs which have the phonetic *kou*/*KU 句, "hook, curve."

Needham's second example, B56, is almost unique in that it has no apparent relation to neighboring sections and there is not a single important word found elsewhere in the Mohist Canon. It is of interest as an especially vivid reminder of the extent to which the understanding of classical Chinese depends on context. We need not criticize Needham's interpretation, because it depends on two purely conjectural emendations. There are at least two rival interpretations which are plausible, utterly different, and require no emendation at all.

PROPOSITION B56

荊之大, 其沈淺也. 說在具.

荊. 沈荊之具也, 則沈淺非荊淺也. [若易五之一].

Solution 1 (tentatively proposed by A. C. G., postulating that the proposition is directed against a presumed sophism to the effect that the difficulty of submerging a bramble, which is traditionally the lightest of woods [Morohashi 30940 sense 3], is the effect and not the cause of its wide extension over the surface of the water).

Canon: The extension of the bramble is because its sub-mergence is shallow. Explained by: the tool.

Explanation: In the case of the tool which submerges the bramble, that its submergence is shallow is not because the bramble is shallow.

TEXTUAL NOTES

We ignore the phrase beginning *jo*, "like," because such phrases are so often misplaced glosses (cf. textual notes to B16 and A48). The word 易 suggests that it belongs to the next proposition, B57, in which the word recurs. The character 具 in the Explanation is corrupted to 貝 in the Taoist Patrology edition, but the 具 confirmed by the Canon is common to most other early editions, as noted in the collation of Wu Yü-chiang.

Solution 2 (Wu Yü-chiang, Liu Ts'un-yan):

Canon: The greatness of [the state] Ching is the weakness of the Shen which belongs to it. Explained by: the tool.

Explanation: If Shen is the tool of Ching, then Shen being weak is not Ching being weak. It is as though it were exchange-able for one-fifth [of Ching].

TEXTUAL NOTES

Kao Heng takes the same position but gilds the lily by emending 具 to 有, "possession." Kao, Wu, and Liu all produce evidence for the relations between Ching (= Ch'u 楚) and Shen, and for 淺 used to mean the weakness of a state.

The uncertainty of both solutions reflects a fundamental problem facing anyone who explores Chinese science. The most interesting documents, which Needham has always shown a sure instinct for smelling out, tend to be off any of the beaten tracks of Chinese literature. Classical Chinese, even more than other languages, can be read with assurance only inside a firm context. The explorer of byways is always looking for related documents, for a subsidiary tradition within which he can find his bearings. Even the present series of eight propositions provides interrelations within which one can begin to become confident about how words are being used, even though optics was never again discussed with this degree of sophistication until the Sung dynasty. In the case of B56, which has no recognized context, at present nothing can be profitably decided. Yet this proposition could become intelligible tomorrow if someone were to chance on a parallel hitherto unnoticed in a contemporary document.

Appendix B. Shen Kua on the Inversion of the Shadow
We have several times referred to the passage on the inversion of the shadow (item 44) in the *Meng hsi pi-t'an* (Dream Creek essays, begun 1088) of Shen Kua. We take the opportunity to provide a translation which deals consistently with the five instances of the crucial word *ai* 礙, "obstruct, obstruction," variously treated in the version of Needham (IV.1, 97):

That things reflected in a burning-mirror are all inverted is because there is an obstruction in the intervening space. Mathematicians call this "the paradigm of contraposition 格術." It is like the way an oar is swept backward and forward by a rower, because the oarlock serves as an obstruction to the movement. When a kite is flying in the sky its shadow shifts in the same direction as the kite. Should the shadow be constricted by a hole in an intervening window, shadow and bird go in opposite directions. If the kite goes east the shadow goes west; if the kite goes west the shadow goes east. It is also like the shadows of pagodas seen through holes in windows. They are constricted by the intervening windows, and likewise all hang upside down, just as in the burning-mirror.

The surface of the burning-mirror is concave. If a finger is reflected close up it will be upright. When it is gradually withdrawn nothing is seen. When withdrawn still further it will be inverted. The place where it is invisible is just like the hole in a window or the oarlock. It obstructs the shadow waist-drum-wise. Base and tip are contraposed 本末相格, so that the same thing happens as with an oar being swept backward and forward. Thus when you lift your hand its shadow goes down, and when you lower your hand its shadow rises. This is the aspect which is visible.

(Note in text: The burning-mirror's surface is concave. When it is exposed to the sun the light all concentrates toward the interior. One or two inches away from the mirror the light concentrates into a spot the size of a seed, and something put there catches fire. This then is the place where the "waist drum" is narrowest.)

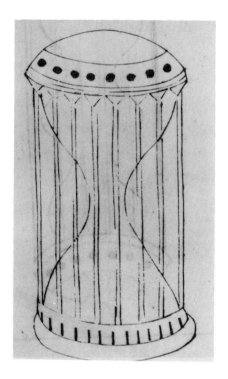

Fig. 6.6. Waist drum, from Ch'en Yang's 陳暘 musical compendium *Yueh shu* 樂書 (presented to the throne in 1104), Canton reprint of 1876, *127*: 10a. Courtesy of the Harvard-Yenching Institute.

It is not only of other things that this is so, it is the same with men. There are few whose vision is not obstructed by things which intervene. At best profit and harm change places and right and wrong are reversed. At worst we treat ourselves as though we were things and things as though they were ourselves. Unless we try to get rid of the obstructions, there is little hope that our vision will not be upside down.

(Note in text: The Yu-yang Miscellany says that when inverted by the sea [?] the shadow of a pagoda turns upside down 海翻則塔影倒, but this is nonsense. That a shadow turns upside down when it enters an opening in a window is its constant principle.)

The passage to which Shen Kua alludes (*Yu-yang tsa tsu*, *4:* 7a5) is obscure, and it is not clear how Shen understands it, although it seems to refer to mirages: "The shadow of the

East Market pagoda suddenly turned upside down. An old man said: 'It is like this when the shadow at sea [mirage?] is inverted 海影翻則如此.' "

Thus Shen explains the inversion by an invisible obstruction surrounding the focal point of the burning-mirror, which he likens to the binding of an oar to its fulcrum and to the surface surrounding the pinhole in the camera obscura (every reader of Chinese novels knows how easily and often snoopers made small holes in paper window panels). This assimilation of burning-mirror to camera obscura implies that, like the Mohist, Shen assumes that the light concentrates at the focal point before reaching the mirror. Indeed, if he had understood that the light is reflected from the curved surface, its concentration at the center would have been intelligible without postulating an unseen obstruction. Unlike the Mohist, Shen notices the disappearance of the image between the center of curvature and the focal point. No ray theory is implied in his conception of the shadow as being "constricted 束" and the light as "concentrated 聚." His geometrical picture is of two cones of shadow meeting at the apexes in the form of a waist drum, and he explicitly distinguishes the unseen cones which he postulates from the "aspect which is visible."

Finally, we may notice that Shen Kua knows the inversion in the concave mirror by experiment with burning-mirrors, but in the camera obscura only by observation of adventitious phenomena.

Bibliography

CLASSICAL TEXTS
Citations in this section are by chüan, page, and line number unless otherwise indicated.

Chia-tzu hsin shu 賈子新書
Ssu pu ts'ung k'an 四部叢刊 ed.

Chuang-tzu 莊子
Harvard-Yenching Institute Sinological Index Series ed. (cited by *p'ien* and line number) for text, *Ssu pu ts'ung k'an* ed. for commentaries.

Hsun-tzu 荀子
Harvard-Yenching Institute Sinological Index Series ed. (cited by *p'ien* and line number).

Kuan-tzu 管子
Ssu pu ts'ung k'an ed.

Lieh-tzu 列子
Ssu pu ts'ung k'an ed.

Meng hsi pi t'an 夢溪筆談
Meng hsi pi t'an chiao cheng 校證 of Hu Tao-ching 胡道靜. References are to item number.

Mo-tzu 墨子
Harvard-Yenching Institute Sinological Index Series ed. (cited by *p'ien* and line number).

Shih chi 史記
Shih chi hui chu k'ao-cheng 會注考證 of Takikawa Kametarō 瀧川龜太郎.

Yu-yang tsa tsu 酉陽雜組
Ssu pu ts'ung k'an ed.

COMMENTARIES, STUDIES, AND REFERENCE WORKS

Chan Chien-feng 詹劍峰
Mo-chia ti hsing-shih lo-chi 墨家的形式邏輯 (The formal logic of the Mohists). Wuhan, 1956.

Chang Ch'i-huang 張其煌
Mo-ching t'ung chieh 墨經通解 (Comprehensive explication of the Mohist Canon). Taipei, 1960.

Chang Ch'un-i 張純一
Mo-tzu chi chieh 墨子集解 (The *Mo-tzu*, with collected explications). Revised ed., Shanghai, 1936.

Ch'en Kuo-fu 陳國符
Tao tsang yuan-liu k'ao 道藏源流考 (Researches in the history of the Taoist Patrologies). Second ed., 2 vols., Peking, 1963.

Ch'en Yuan 陳垣
"Shih hui chü li 史諱舉例" (Examples of avoidance of name taboos in history), *Yen-ching hsueh-pao* 燕京學報, 1928, *4*: 537–650.

Ch'ien Lin-chao 錢臨照
"Shih Mo-ching chung kuang-hsueh li-hsueh chu t'iao 釋墨經中光學力學諸條" (An explication of the optical and dynamical propositions in the Mohist Canon), in *Li Shih-tseng hsien-sheng liu-shih-sui chi-nien lun-wen chi* 李石曾先生

六十歲紀念論文集 (Studies presented to Mr. Li Shih-tseng on his sixtieth birthday). Kunming, 1940. Not available.

———

"Lun Mo-ching chung ti kuan-yü hsing-hsueh li-hsueh yü kuang-hsueh ti chih-shih 論墨經中的关于形學力學与光學的知識" (On knowledge of geometry, mechanics, and optics in the Mohist Canon), *Wu-li t'ung pao* 物理通報, Vol. 1, No. 3. Not available.

———

"Yang sui 陽燧" (Burning-mirrors), *Wen-wu ts'an-k'ao tzu-liao* 文物參考資料, 1958, no. 95, pp. 28–30. A reference to a fourth article in Hung Chen-huan (p. 2, n. 2) is incorrect, and we have not been able to locate it.

Chu Ch'i-feng 朱起鳳
Tz'u t'ung 辭通 (Dictionary of variant forms). 2 vols., Shanghai, 1934, reprinted Taipei, 1960.

Fang K'ao-po 方考博
"Mo-ching chung ti shih k'ung kai-nien yü kuang-hsueh li-lun 墨經中的時空概念與光學理論" (The concepts of time and space and optical theory in the Mohist Canon), *Lan-chou ta-hsueh hsueh-pao* 蘭州大學學報, 1957, no. 1. Not seen. Cited by Hung from a copy corrected by the author.

Feng Han-ch'u 馮涵初
Kuang-hsueh shu Mo 光學述墨 (An elucidation of the optics in the *Mo-tzu*). Place unknown, 1910. Not seen; listed in Yen Ling-feng, *1.* 2: 30.

Forke, Alfred
Mê Ti des Sozialethikers und seiner Schüler philosophische Werke. Berlin, 1922.

Graham, A. C.
"The Logic of the Mohist Hsiao-ch'ü," *T'oung Pao* (Leiden), 1964, *51*. 1: 1–54.

———

"The Grammar of the Mohist Dialectical Chapters," in *A Symposium on Chinese Grammar* (ed. Inga-Lill Hansson; Scandinavian Institute of Asian Studies Monograph Series, VI; Lund, 1971), pp. 55–141.

Hung Chen-huan 洪震寰
" 'Mo ching' kuang-hsueh pa t'iao li shuo '墨經' 光學八條釐說" (A new explanation of the eight optical propositions of the Mohist Canon), *K'o-hsueh-shih chi-k'an* 科學史集刊, 1962, *4:* 1–40.

———

"Wo kuo ku-tai ti ch'iu mien ching chi ch'i-t'a 我國古代的球面鏡及其他" (Ancient Chinese spherical mirrors and related topics), *Hang-chou ta-hsueh hsueh-pao* 杭州大學學報, 1960, no. 1 (Physics issue). Not seen; listed in the preceding article, p. 32.

Kao Heng 高亨
Mo ching chiao ch'üan 墨經校詮 (Critical edition and explication of the Mohist Canon). Peking, 1958, 1962. First draft completed 1944.

Karlgren, Bernhard
The Book of Odes. Stockholm, 1950.

Kuo Ch'ing-fan 郭慶藩
Chuang-tzu chi shih 莊子集釋 (The *Chuang-tzu,* with collected explications). 2 vols., Taipei, 1958.

Li Yü-shu 李漁叔
Mo pien hsin chu 墨辯新注 (New annotations to the Mohist dialectical writings). Taipei, 1968.

Liang Ch'i-ch'ao 梁啓超
Mo-ching chiao-shih 墨經校釋 (Critical edition and explication of the Mohist Canon). Shanghai, 1920.

Liang Heng-hsin 梁恒心
"Wo kuo ku-tai k'o-hsueh-chia tui she-ying chi-pen chih-shih ti kung-hsien 我國古代科學家對攝影基本知識的貢獻" (The contributions of our ancient scientists toward a basic understanding of photography), *Ta chung she-ying* 大众攝影, 1960, no. 3, pp. 17–19.

Liu Ts'un-yan [Ts'un-jen] 柳存仁
"Mo-ching chien i 墨經箋疑" (Commentary on problematic passages of the Mohist Canon), *Hsin Ya hsueh pao* 新亞學報, 1964, 6. 1: 45–140; 1965, 7. 1: 1–134.

Lu Ta-tung 魯大東
Mo pien hsin chu 墨辯新注 (New annotations to the Mohist dialectical writings). Shanghai, 1936.

Luan Tiao-fu 欒調甫
Mo-tzu yen-chiu lun-wen chi 墨子研究論文集 (Collected research papers on the *Mo-tzu*). Peking, 1957. Articles originally published between 1922 and 1932 on all aspects of the *Mo-tzu.*

Mei, Y. P.
The Ethical and Political Works of Mo-tse. London, 1929.

Morohashi Tetsuji 諸橋轍次
Dai Kan-Wa jiten 大漢和辭典 (Unabridged Chinese-Japanese dictionary). 13 vols., Tokyo, 1955–1960. Citations are to character and compound number (for example, 1234..23) or character and definition number (for example, 1234 sense 3).

Needham, Joseph
Science and Civilisation in China. 7 vols. projected, Cambridge, England, 1954–.

Pi Yuan 畢沅
Mo-tzu 墨子. *Ssu pu pei yao* 四部備要 ed. This commentary (1783) and that of Sun I-jang (1894) listed in the following entry remain basic.

Sun I-jang 孫詒讓
Mo-tzu chien ku 墨子閒詁 (The *Mo-tzu*, with ameliorative and explanatory annotation). Peking, 1954.

T'an Chieh-fu 譚戒甫
"Mo-ching kuang-hsueh 墨經光學" (The optics of the Mohist Canon), *Tung-fang tsa-chih* 東方雜誌, 1933, *30.* 13: 154–168.

———
Mo-ching i chieh 墨經易解 (The Mohist Canon simply explained). Shanghai, 1935.

———
Mo pien fa wei 墨辯發微 ("Revelation of the hidden" in the Mohist dialectical writings). Peking, 1958; corrected ed., 1964. T'an's article is an explication of the eight optical propositions, with only a short introduction. We cite the 1964 book.

Ts'ao Yao-hsiang 曹耀湘
Mo-tzu chien 墨子箋 (Critical edition of the *Mo-tzu*). Changsha, 1906.

Ts'en Chung-mien 岑仲勉
Mo-tzu ch'eng shou ko p'ien chien chu 墨子城守各篇簡注 (The *Mo-tzu* chapters on the defense of walled cities, concisely annotated). Peking, 1958.

Tuan Wei-i 段維毅
Ku chuan wen ta tzu-tien 古篆文大字典 (Unabridged dictionary of ancient seal script). Taichung, 1965.

Wang Tien-chi 汪奠基
Chung-kuo lo-chi ssu-hsiang shih-liao fen-hsi 中國邏輯思想史料分析 (A documentary analysis of the history of logical thought in China). First collection, Peking, 1961. The section on the Mohists includes a proposition-by-proposition interpretation of the Mohist Canon (pp. 286–371), unfortunately offhand and derivative.

Watanabe Takashi 渡邊卓
"Bokka no shūdan to sono shisō 墨家の集團とその思想" (The Mohists' organization and their thought), *Shigaku zasshi* 史學雜誌, 1961, *70:* 1198–1231, 1351–1385.

"Bokka no heigikōsho ni tsuite 墨家の兵技巧書について" (About the Mohists' writings on the military arts), *Tōkyō shinagaku hō* 東京支那學報, 1957, *3:* 1–19. Important for the composition and dating of the military chapters.

Watson, Burton
Mo-tzu. Basic Writings. New York and London, 1963.

Wu Yü-chiang 吳毓江
Mo-tzu chiao chu 墨子校注 (The *Mo-tzu,* critically annotated). Chungking, 1944.

Yabuuti Kiyosi 藪內清
Bokushi 墨子 (The *Mo-tzu*). Printed together with Kakimura Takashi 柿村峻, *Kampishi* 韓非子 (The *Han-fei-tzu*) in *Chūgoku koten bungaku taikei* 中國古典文學大系 (Classics of Chinese literature), vol. V. Tokyo, 1968.

Yang Po-chün 楊伯峻
Lieh-tzu chi shih 列子集釋 (The *Lieh-tzu,* with collected explications). Shanghai, 1958; reprint, Hong Kong, 1965.

Yen Ling-feng 嚴靈峯
"*Mo-tzu chih chien shu-mu* 墨子知見書目 " (A *Mo-tzu* bibliography), *Kuo-li chung-yang t'u-shu-kuan kuan k'an* 國立中央圖書館館刊, 1967, n. s., *1.* 2: 15–38; 1968, n. s., *1.* 3: 21–41.
    A carefully annotated list of editions, commentaries, and general studies published in book form.

7

Elixir Plants: The Ch'un-yang Lü Chen-jen yao shih chih 純陽呂眞人藥石製 (Pharmaceutical Manual of the Adept Lü Ch'un-yang)

Ho Peng Yoke, Beda Lim, and Francis Morsingh

In former times men sought the strangest end—
To make from metals base a precious stuff,
Or live eternal, and thus death transcend
When this our earthly span is life enough.
On plants, on metals they did much research,
And numerous other substances they tried,
Shunning distractions in their lonely perch;
But gold they have not made, and they have died.
You, our teacher and friend, have chosen right:
You know gold comes from arduous industry.
This law you passed on to your acolyte—
In learning lies our surest alchemy.
    True gold you wrought from years of work and pain,
    True immortality, unsought, you gain.

Vegetables and other plants are prominent among the elixir substances ingested by the adepts depicted in the *Lieh hsien chuan* 列仙傳 (Lives of famous *hsien* or Taoist immortals), a work that has been attributed to Liu Hsiang (ca. 50 B.C.), but was in fact written by an unknown Taoist who lived between the second and early fourth centuries A.D. However, the attention of the early Chinese alchemists, from the time of Han Wu-ti (r. 140–87 B.C.) to the second half of the T'ang dynasty in the ninth and tenth centuries, was focused mainly on metals and minerals. Because of the many instances of elixir poisoning the alchemists became increasingly cautious in the use of minerals. There were those who tried to find antidotes to the toxicity of minerals or to exercise great care in their choice, attributing the disastrous result to mistakes made in the identification of the genuine material. Many turned to physiological meditational alchemy, hoping to attain longevity and physical immortality through meditation and proper control of respiration. Some adepts, however, searched for immortality in the vegetable kingdom. But the name of any great alchemist of plants and vegetables is still unknown, and alchemical texts on elixir plants are relatively scarce compared to those on metals and minerals. Among the former is the *Ch'un-yang Lü Chen-jen yao shih chih,* the subject of our present study.

The text of the *Ch'un-yang Lü Chen-jen yao shih chih* gives no indication whatsoever as to the name of its author and the date when it was written.[1] As we cannot find out much about its anonymous author, we shall try to establish its approximate date. The adept Lü Ch'un-yang is better known as Lü Tung-pin 呂洞賓 or Lü Yen 巖 and is a leading member of the so-called Eight Immortals (*pa hsien* 八仙) in popular Chinese folklore. We are not very certain about his dates. His year of birth has been variously given as A.D. 755 and 796, and some sources suggest that he flourished during the ninth and tenth centuries. Hence we may begin with the most liberal guess that the text was not written before the second half of the eighth century. We also find that the title of this book is included in the Taoist Patrology of 1444, the *Cheng-t'ung tao tsang* 正統道藏, indicating that it must have existed before the middle of the fifteenth century. We shall next proceed to narrow down the range as much as we can.

A comparison of the plant names in the text with those found in Chinese pharmacopoeias gives some indication of a rather late date. Among its sixty-one plant names, twelve find no exact equivalents in the pharmacopoeias, and twenty-eight occur in pharmacopoeias written before the ninth century: fifteen in the *Shen-nung pen-ts'ao ching* 神農本草經 (first century A. D.), one in Wu P'u's 吳普 (ca. A. D. 225) writings, seven in T'ao Hung-

---

[1] This is discussed in greater detail in Chen T'ieh-fan and Ho Peng Yoke, "Lun *Ch'un-yang Lü Chen-jen yao shih chih* ti chu-ch'eng shih-tai 論純陽呂真人藥石製的著成時代," *Journal of Oriental Studies* (Hong Kong), 1971, *9*: 181-229.

Table 7.1. Occurrences of Plant Names in Pharmacopoeias

Name of Pharmacopoeia	Number of Instances Where Plant Name Is First Mentioned
Shu pen-ts'ao 蜀本草 (tenth century)	1
K'ai-pao pen-ts'ao 開寶本草 (ca. 970)	3
Jih-hua-tzu 日華子 (tenth century)	3
Chia-yu pen-ts'ao 嘉祐本草 (ca. 1057)	1
Pen-ts'ao t'u-ching 本草圖經 (ca. 1070)	6
The lost Sung pharmacopoeia of T'u-hsiu Chen-chün 土宿眞君	1
The pharmacopoeia of Li Kao 李杲 (early thirteenth century)	1
Keng hsin yü ts'e 庚辛玉冊 (ca. 1430)	1
Pen-ts'ao kang mu 本草綱目 (1596)	4

ching's 陶弘景 (fifth century) works, four in Su Ching's 蘇敬 *T'u ching pen-ts'ao* 圖經本草 (659), and one in Meng Shen's 孟詵 pharmacopoeias (ca. 670). The remaining twenty-one plant names are found mentioned for the first time in pharmacopoeias of much later date, as shown in Table 7.1. The late occurrence of some of the plant names in the text suggests that its date might be much closer to its lowest limit, 1444, than its early limit, the eighth century.

The text consists of sixty-nine verses, one of which (XLVI) is missing and one other (XLV) incomplete. An investigation of the prosody in these verses has thrown some light on the approximate date of their composition. With only two exceptions all the verses are of the type known as "seven-word broken-off lines" (*ch'i yen chueh chü* 七言絕句); that is, each stanza consists of four lines of seven characters each. The other two (VIII and XIX) are the so-called five-word broken-off lines (*wu yen chueh chü* 五言絕句). If the *Kuang yun* 廣韻 dictionary of sounds (A.D. 1011) is taken as a criterion then only fifteen stanzas are found to rhyme perfectly (III, V, VI, IX, XXII, XXVI, XXXIV, XXXVI, XXXIX, L, LII, LIV, and LVIII). An astonishing result appears when we apply the yardstick of similar dictionaries of later dates. Twelve more stanzas will rhyme if we take the *P'ing shui yun* 平水韻 of Liu Yuan 劉淵 (1252) (I, II, IV, XIV, XV, XXV, XXXVII, XLVIII, LI, LXII, LXIII, and LXVIII). Another seventeen are in rhyme by the standards of Chou Te-ch'ing's 周德清 *Chung yuan yin yun* 中原音韻 of 1324 (XIII, XVIII, XIX, XX, XXI, XXIX, XXXII, XXXV, XXXVI, XL, XLIII, XLVII, XLIX, LVII, LIX, LXI, and LXV). To cap it all, all the stanzas are found to rhyme on the basis of the *Wu fang yuan yin* 五方元音 of Fan T'eng-feng 樊騰鳳 (1624–1672). Now, if all the stanzas were meant to rhyme then they could not be written before the *Chung yuan yin yun*, that is, before 1324. This late date is supported by the plant names given in the text.

All the verses in the book cannot by any means be regarded as beautiful poetry. They could not have come from the pen of a distinguished poet or scholar, but rather from that of a very ordinary literatus. The verses themselves also indicate the writer's keen interest in and knowledge of plant elixirs. This aspirant to immortality could well have been a Taoist adept who spent much of his time roaming among the mountains and the countryside making his botanical studies. Hence it is quite probable that the *Ch'un-yang Lü Chen-jen yao shih chih* was composed by an anonymous Taoist adept some time between the years 1324 and 1443.[2]

There is evidence in the Taoist Patrology that this book belonged to a tradition. The *P'eng-lai shan hsi tsao huan tan ko* 蓬萊山西竈還丹歌 (Vol. 592), for instance, is a collection of 172 verses, each about a different vegetable immortality substance, each prefaced by a short article giving common and literary

[2] It so happened that this fell in the period when the first illustrated botanical monograph on edible wild plants was written in China, the Famine Relief Pharmacopoeia or *Chiu huang pen-ts'ao* 救荒本草 by Chu Shu 朱橚 (1406). This work contains 414 plant names, 276 of which are not contained in the pharmacopoeias. No less than thirty percent of the plant names given in the *Ch'un-yang Lü Chen-jen yao shih chih* are also given in this work, including some as obscure as *yang chiao miao* (stanza XLVI). This was to be followed in the next two centuries by three further works of similar nature, namely Wang P'an's 王磐 *Yeh-ts'ai p'u* 野菜譜, Chou Lü-ching's 周履靖 *Ju-ts'ao pien* 茹草篇, and lastly Pao Shan's 鮑山 *Yeh-ts'ai po lu* 野菜博錄 (1622).

names and descriptions. It is attributed to one Huang Hsuan-
chung 黃玄鍾 and carries a preface addressed to Emperor Wu of
the Han dynasty (137–87 B.C.), but since the poems are in
"five-word broken-off lines" it could not possibly be earlier
than the T'ang. Another text which cannot be dated, the
Hsien-yuan huang-ti shui ching yao fa 軒轅黃帝水經藥法 (vol. 597),
a series of 32 methods of bringing minerals into solution, generally using herbs and vinegar, incorporates a list of identifications
of various "dragon sprouts" which corresponds on the whole to
those in the work we are translating.

Many of the stanzas in the text refer to the changing of base
metals into gold. Any process that would give a yellow goldlike
coating to the former could have been included among the
various techniques of the Chinese aurifactors. We know that
nature abounds in organic pigments that have a rich golden or
yellow color. Many of these pigments are masked by the green
chlorophyll present in plants, but by careful separation they
can be isolated in a pure form. The group of pigments under the
general heading carotenoids (Willstatter, *Annalen der Chemie*,
1907, *355*: 1) contains a variety of yellow and golden pigments.
Lycopene obtained from the golden flowers of the marigold
(Zechmeister, *Zeitschrift für physiologische Chemie*, 1932, *208*: 26);
carotene, a yellow coloring matter from autumn leaves (Karrer,
Helvetica Chimica Acta, 1933, *16*: 641); xanthophylls from green
leaves, maize, and egg yolk (Kuhn, Winterstein, and Lederer,
Z. physiol. Chem., 1931, *197*: 141); and kryptoxanthin, the main
pigment of the mandarin orange (Kuhn, Brockmann, *Z.
physiol. Chem.*, 1932, *206*: 41) are but a few examples. In the
process of working up the plant substances mentioned in our
text we cannot easily rule out the possibility that the Chinese
alchemists had succeeded in isolating some goldlike compounds
in a reasonably pure form. However, at the same time we are
mindful of the liberal employment of metaphors and synonyms
in classical Chinese poetry. Gold making was firmly believed
to be a prerequisite and the first step toward the attainment of
physical immortality at the time of the Emperor Han Wu-ti
(r. 140–86 B.C.). It is quite possible that longevity—if not

immortality—was generally alluded to when the anonymous author of our text referred to transmutation.

All the plants in the text are regarded and named as varieties of "Dragon Sprout" (lung-ya 龍芽). It is difficult to guess why they are so designated. Elsewhere the term lung-ya appears in one of two forms, one with the word ya lacking the "grass" radical, meaning "teeth," and one with ya meaning "sprout." Sometimes these two variant forms are interchangeable; for example, both "dragon teeth" (lung-ya 龍牙) and "dragon sprout" have been used as synonyms for tea.[3] However, this is not true in other instances; for example, the "dragon teeth plant" (lung-ya-ts'ao 龍牙草), mentioned in the early twelfth-century pharmacopoeia of K'ou Tsung-shih, has been identified as Verbena officinalis, L.,[4] while the "dragon sprout plant" (lung ya ts'ao 龍芽草) has been regarded by Li Shih-chen as she-han 蛇含, Potentilla kleiniana, Wight.[5] The word lung (dragon) has often been used to refer to the emperor, or to something that is full of excellence, or that is noble, talented, or auspicious, and it has also been applied in many other plant names. Ya (sprout) means a bud, a shoot, or a germ. Hence one may infer the connotation "auspicious plants" from the term "dragon sprout" in our text.

For this special occasion, in dedicating this article in celebration of the seventieth birthday of Dr. Joseph Needham, we would like to interpret "dragon sprout" in another sense. We may read lung (the dragon) as an abbreviated form of lung ma ching shen 龍馬精神, meaning a vigorous old age—like that of a dragon or a horse. In Japan the word for sprout conveys the vision of early spring, when the buds of the trees shoot forth to a new life, and it is used in phrases like me ga deru 芽が出る, meaning "fortune smiles upon you," and omedetō gozaimasu

---

[3] "Dragon teeth" is also the name given to an unusual variety of lichee. See Chih-wu ming shih t'u k'ao ch'ang p'ien, p. 854.

[4] See Chih-wu ming shih t'u k'ao, p. 181, and plant no. 147 in Read's Chinese Medicinal Plants from the "Pen Ts'ao Kang Mu" (hereafter abbreviated as R147 and similarly).

[5] See Chih-wu ming shih t'u k'ao ch'ang p'ien, p. 291, and R442.

お芽出度う御居います, meaning "congratulations!" or "we wish you joy." It is in the spirit of this rather liberal interpretation that we are now rewriting the verse of the *Ch'un-yang Lü Chen-jen yao shih chih* in English.

I. *T'ien-pao lung-ya* 天寶龍芽
(The "Heavenly Precious Dragon Sprout")
ch'ih-ch'in 赤芹

This comes first among the plants:
Handled by predestined men
This, by art and not by chance,
Darkly, in the scholar's den,
Turns base metals into gold.
Metals Five and Minerals Eight—
Easily this plant can mold
And to nobler stuff translate.
Common copper it can change
Into metals fanciful;
But the plant—it's passing strange—
Stays itself as soft as wool.

Ch'ih-ch'in is called *tzu-ch'in* 紫菫 by Su Ching, according to the *Pen-ts'ao kang mu* (hereafter abbreviated as *PTKM*), *26*: 1201.2. According to the *Hsuan-yuan pao tsang lun* the juice extracted from this plant has been used for boiling orpiment and reacting with mercury, with cinnabar, or with each of the so-called Three Yellow (Substances), that is, sulfur, realgar, and orpiment. It has been identified as *Corydalis incisa*, Pers. (R488)

The "Five Metals" are gold, silver, copper, iron, and lead, while the "Eight Minerals" are the Five Metals and the Three Yellow (Substances). In a broader sense the term "Five Metals and Eight Minerals" seems to mean metals and minerals in general. For an explanation of these terms see Ho Ping-yü and Joseph Needham, "Theories of Categories."

II. *Pao-sha lung-ya* 寶砂龍芽
(The "Precious Ore Dragon Sprout")
sang yeh 桑葉

Green the leaves, its flowers white,
Growing wildly in the plains,
Seen in farms, though recondite,
This to high estate attains.
Common people fail to see
That Immortals eat this fruit,
Gaining thus eternity;
That this same plant can transmute
Common copper into gold,
While its nature it retains:
Though the metal it will mold,
Soft as wool the plant remains.

Sang yeh, or mulberry leaves (R605), contain calcium malate, calcium carbonate, invert sugar, pentosan, tannin, carotene, 10 percent ash, vitamin C_2, and cholin, according to the analysis of a sample by Read. It was popular among the early aspirants to immortality, for Ko Hung 葛洪 in the fourth century has described its use as an elixir component more than once (see, for example, Ware, *Alchemy, Medicine, Religion,* p. 190). The fact that one of its synonyms was *shen-hsien-ts'ai* 神仙菜, "vegetable of the immortals," shows that it must have been regularly consumed by adepts (*PTKM, 36*: 1430.2).

III. *Tui-chieh lung-ya* 對節龍芽
(The "Dragon Sprout with Opposed [Leaves in a] Joint")
i-mu 益母

In the wild this plant will grow,
White its flowers, green its leaves,
Men of secret science know
What this humble plant achieves:
Heated and refined by fire,
Frosty powder will result,
Magic strength it will acquire,
Make the scholar's heart exult.
Minerals Eight and Metals Five
Bow before its potency;

Crows that eat of it will thrive
And attain longevity.

I-mu is a synonym for *ch'ung-wei* 充蔚 (see *PTKM 15*: 856.2), which is identified as *Leonurus sibiricus* L., or Siberian mother-wort. Another synonym is *hsia-k'u-ts'ao* 夏枯草, similar to no. LVIII. It is composed of leonurin, leonurinine, essential oil, and fatty oil (R126). It was said by alchemists to react with sulfur, orpiment, and arsenious oxide (for instance, in the *Tan fang ching yuan*, quoted by Li Shih-chen in *PTKM, 15*: 856.2). This plant is listed in *Chiu huang pen-ts'ao* (hereafter abbreviated *CHPT*), *1A*:13b under the name *yü-ch'ou-miao* 蔚臭苗.

IV. *Wei-t'ang lung-ya* 味棠龍芽
(The "Delicious Pear Dragon Sprout")
tu-li-erh 杜梨兒

Growing on the high hill slopes,
This uncultivated plant
Offers scholars many hopes—
Hidden powers it will grant.
When the fruit is ripe and sweet
It may serve our useful ends:
Rice and vinegar will greet
These wild fruits as long lost friends.
Lowly matter it will fix,
And can act on mercury;
Alchemists will, on its tricks,
Spend their time and energy.

Tu-li-erh is very probably *t'ang-li* 棠梨, here called *tu*, "small coarse pear," of which there are two varieties, *Pirus betuliafolia*, Bge. and *Pirus baccata*, L. (R432). Li Shih-chen (*PTKM, 30*: 1271.1) says that the red variety is called *tu* and the white variety *t'ang*, or sometimes the female *tu* and the male *t'ang*, and also the sour variety *tu* and the sweet *t'ang*. As the synonym given here is *wei-t'ang lung-ya*, the "delicious pear dragon sprout," it is likely that the sweet variety is being referred to.

V. *Erh-ch'i lung-ya* 二氣龍芽
(The "Dragon Sprout of the Two Cosmological Pneumata")
shan-ho-yeh 山荷葉

Using only nature's skills,
Backed by Water and by Fire,
Through this plant from far-off hills,
Cinnabar we can acquire.
Lakes and mountain left behind,
Scholar's den is good enough:
Sulfur, mercury, we find
Fixed by this to rarer stuff.
With this plant the Alchemist
Holds success within his view,
Nature's power will assist,
Flawless jade will be his due.

Shan-ho-yeh is a synonym for *kuei-chiu* 鬼臼, a name now applied
to *Diphylleia cymosa* or umbrella leaf (see R520). A T'ang
specimen preserved in the Shōsōin turned out to be the rhizome
of a *hosta* species (Sivin, *Chinese Alchemy*, p. 281 and Asahina,
Shōsōin yakubutsu, p. 177). Of course we cannot say definitely
what it referred to in the fourteenth and fifteenth centuries.
According to Li Shih-chen its roots were used by the alchemists
to "fix" sulfur, realgar, orpiment, cinnabar, and mercury.
He also quotes the *Tan fang ching yuan*, saying that it "fixes"
cinnabar and mercury (*PTKM, 17B:* 985.2).

VI. *Wu-feng lung-ya* 五鳳龍芽
(The "Five Phoenixes Dragon Sprout")
kuan-chung 菅仲

Deep within the mountain streams,
"Phoenixes Five" is their name;
These the Alchemist esteems,
And their lushness will acclaim.
Harvesting the leaves is done
Only in the seasons blest:
Fifth moon or mid-autumn sun—

Under these the task is best.
Strange the name and strange the shape—
This is no prosaic thing—
For the leaves distinctly ape
Phoenix tail and phoenix wing.

Kuan-chung, which reminds us of the celebrated statesman of Ch'i State during the Spring-and-Autumn Period, is a misnomer for its homophon *kuan-chung* 貫衆, *Aspidium falcatum* or wood fern, according to Li Shih-chen (*PTKM, 12B:* 747.2). Wu P'u 吳普 explains that its leaves were plucked during the fifth lunar month, while its roots were gathered in the eighth month, that is, in mid-autumn. Analysis shows that its roots contain β-filicic acid, uspidinol, flavaspidic acid, albaspidin, filicitannic acid, and filicin (R800A). One of its synonyms is *feng-wei-ts'ao* 鳳尾草, referring to its resemblance to the tail of a phoenix.

VII. *T'ien-jen lung-ya* 天奴龍芽
(The "Celestial Blade Dragon Sprout")
ch'ang-p'u 菖蒲

In the early stage of life
Like a sword this will be seen;
White roots on the ground run rife,
And its leaves are dense and green.
Only in the fourth moon may
Gathering of this crop be done,
And the cunning sages say
Other seasons we must shun.
Strange the virtue of this herb:
Add realgar, agitate—
Nature's way it will disturb,
Brighter gold precipitate.

Ch'ang-p'u is *Acorus gramineus,* Ait., whose roots contain asarone, $C_{12}H_{16}O_3$, 0.8 percent essential oil, palmitic acid, sesquiterpene, and phenols (R704). It appears to be a plant sought after by the early aspirants of immortality. Ko Hung in the fourth century described this as the food of the immortal Han Chung (*PTKM, 19:* 1063.2 ff. and *Pao p'u tzu nei p'ien,* ch. 11, in Ware, p. 209). This is included among the edible plants in the *CHPT, 1B:* 29a.

VIII. *Ti-chin lung-ya* 地錦龍芽
(The "Embroidered Carpet Dragon Sprout")
i-pan 衣班

Deep within the mountain range
You will find this mystic plant,
And its properties are strange—
Hidden powers it will grant.
Flowers red, and red its leaves,
And the leaves are round like coins;
Strange effects the sage achieves—
To the humble leaves he joins
Some realgar—the result:
Medicine's made successfully;
Eaten as a sacred cult
This will give longevity.

The name *i-pan* has not been identified. However, both the context and the name *Ti-chin lung-ya* 地錦龍芽, the "embroidered carpet dragon sprout," give strong indication that this is *ti-chin* 地錦, *Euphorbia humifusa* Willd., or thyme-leaf spurge (R325). Li Shih-chen says that this was gathered by the alchemists in the autumn months and used for boiling together with orpiment, realgar, cinnabar, or sulfur.

IX. *Chin-so lung-ya* 錦鑠龍芽
(The "Inlaid Lock Dragon Sprout")
hsu-tuan 續斷

That it's fragile, one admits;
This the artisan can see,
Nothing does he know of its
Pharmaceutic formulae.

As with other mystic plants,
Magic powers it will bring:
Sal ammoniac it enchants,
Turns it to a precious thing.
Chief device in all its range
Bears upon identities:
Common copper it can change
Into nobler entities.

Hsu-tuan is *Lamium album,* L. or white nettle. Its roots contain stachyase, glucoside, and saponin (R124). It has been used by bone-setters, and hence its name (literally, "break-joiner"; see *PTKM, 15:* 867.1).

X. *Kan-lu lung-ya* 甘露龍芽
(The "Sweet Dew Dragon Sprout")
kan-ts'ao 甘草

Highly potent are these leaves,
Scanty on a single stalk;
In their strength the sage believes,
Of their magic he will talk.
Yellow Emperor honored them
With the name "State Elders"—yes,
Great Work does include this gem—
Roots and sprouts that men will bless
And the Alchemist reveres.
Arsenic the leaves subdue,
Frosty, silvery it appears,
Turning into something new.

Kan-ts'ao is *Glycyrrhiza glabra,* L. or licorice. Its roots contain glycyrrhizin, glucose, protein, mannite, asparagine, resin, Ca, Mg, NH_4, urease, glucuronic acid, and saccharase (R391). One of its synonyms is *kuo-lao* 國老, "state elder." T'ao Hung-ching explains this synonym by the fact that this substance could be used together with any other drug with beneficial results; ancillary drugs were given bureaucratic titles in Chinese medicine (Sivin, p. 153, n. 14; *PTKM, 12A:* 717.1).

XI. *Chin-mei lung-ya* 金美龍芽
(The "Golden Beauty Dragon Sprout")
yang-t'i 羊蹄

Purple leaves adorn this breed,
Scarlet are the roots it sports;
Prism-shaped we find the seed:
To this plant the sage resorts.
Moist the places where it sprouts:
Summer sees it in its prime,
But midwinter, thereabouts,
Is its favored harvest time.
Here's a plant the sage salutes;
It reacts on calomel,
Copper, silver, it transmutes—
Such the nature of its spell.

Yang-t'i is *Rumex orispus*, L. or yellow dock. Its root contains
emodin, chrysophanic acid, anthraquinone, tannin, essential
oil, and calcium oxalate, while its leaves contain emodin and
brassidic acid (R584). K'ou Tsung-shih says that the alchemists
use this to "fix" lead and mercury. (*Ch'ung hsiu Cheng-ho ching
shih cheng lei pei yung pen-ts'ao*, hereafter *CLPT, 11:* 267.1).

XII. *Mi-wu lung-ya* 覓烏龍芽
(The "Crow-Attracting Dragon Sprout")
ying-t'ao 櫻桃

Dragonlike this tree suspends;
From the heights of cliffs remote
To the lowland it extends.
Of the leaves take special note:
Fresh as spring do these appear
As though hillside-village grown.
Kept in cellars until sear,
Healing powers soon are shown,
Specially so with larger fruits.
Sal ammoniac gives to it
Certain wondrous attributes:
Jadelike luster exquisite.

Ying-t'ao is *Prunus pseudo-cerasus,* Lindl., or cherry. The fruit contains 16.5 percent carbohydrates, 1 percent protein, 0.8 percent fat, 0.6 percent ash (R449). Li Shih-chen mentions that it attracts birds when ripe (*PTKM,* 30: 1298.1). See also *CHPT,* 2B: 10a.

XIII. *Chin-ssu lung-ya* 金絲龍芽
(The "Golden Thread Dragon Sprout")
t'u-ssu 兎絲

From the branches hang these strands,
Glistening like golden threads;
Placed in knowledgeable hands
Virtue from this foliage spreads:
Seventh moon its harvest time;
Boiled in water it will turn
Sulfur into stuff sublime—
This it pays us all to learn.
One of its gifts will amaze—
Magic transport it incites:
Eat it for a hundred days,
And you fly to far-off heights.

T'u-ssu is probably *t'u-ssu-tzu* 菟絲子, *Cuscuta japonica,* Choisy or dodder. Its seeds contain resin and sugars (R156). This has been used as an elixir substance before the time of Ko Hung, who mentions it in his *Pao p'u tzu nei p'ien* (e.g. Ware, p. 85). Ko Hung also says that its new roots look like rabbits and the juice extracted from them makes a drink with which to swallow elixirs (*PTKM, 18A:* 1002.2).

XIV. *Wu-yu lung-ya* 無憂龍芽
(The "Without-Anxiety Dragon Sprout")
hsuan-ts'ao 萱草

Named "Without Anxiety"
This is an unusual herb.
Everyday society
Knows not of its gifts superb,
Though Immortals know its tricks.

As the common people strive
To determine how it ticks,
Even in their earnest drive
To the mystic discipline
Efforts atrophy and die:
Time is passing fast within
Sacred island of P'êng-lai.

Hsūan-ts'ao is *Hemerocallis fulva*, L. or yellow day lily. Its fruits contain vitamins A, B, and C, and its roots asparagine and colchicine (R679). One of its synonyms is *wang-yu* 忘憂, "forgetting all cares and sorrows," (*PTKM, 16:* 901.1), and hence its name, "without-anxiety dragon sprout." See also *CHPT, 1A:* 6b.

XV. *Sui-yen lung-ya* 碎焰龍芽
(The "Flame-Crushing Dragon Sprout")
hu-chai 護宅

Small the courtyards where it's found;
Grown in pots this humble tree
Harbors gifts that will astound.
How do wise men seek its key?
Sedulous experiment,
Through which path they find this fact:
With this fragile element
Sal ammoniac will react,
Copper, iron, turn to white
When the metals this confronts.
Human ailments it can fight—
Common cough it stops at once.

Hu-chai, literally "guardian of the house," considered with the arcane name "Flame-Crushing Dragon Sprout," is very probably identifiable with *hu-huo* 護火 "guardian against fire," which is a synonym of *ching t'ien* 景天, *Sedum alboroseum*, Bak., or stonecrop. Its leaves contain malic acid and its flowers calcium malate (R471). T'ao Hung-ching says that this plant was cultivated in pots and placed on the roof so as to avert the outbreak of fire. (See *PTKM, 20:* 1079.1.) According to *Jih-hua-tzu* 日華子, it was used for roasting cinnabar (*PTKM, 20:* 1079.2).

XVI. *Pai-hsüeh lung-ya* 白雪龍芽
(The "White Snow Dragon Sprout")
t'u-ch'uang-hua 禿瘡花

Leaves of this are bluish tint,
And its flowers are white as snow;
To the dunce it gives no hint
Of the gifts it can bestow.
By its strength it can affix
Red realgar, and react
With this prince of arsenics:
Well the sages know this fact.
When Immortals contemplate
Savage plants, they testify:
Eat this magic herb, and straight
To the heavens you will fly.

T'u-ch'uang-hua is a common name for *t'ung-ch'uan-ts'ao* 通泉草, otherwise called *ch'ang-sheng-ts'ao* 長生草, according to the *Keng hsin yü ts'e* 庚辛玉冊 (see *Chih-wu-hsüeh ta-tz'u-tien*, p. 941.1). Identified as *Mazus rugosus*, Lour., its leaves can serve as a vegetable.

XVII. *Wu-hsin lung-ya* 無心龍芽
(The "Undesigning Dragon Sprout")
pan-hsia 半夏

Growing from a bulbous stem,
Leaves in sheaths are clearly seen;
In the fifth moon will this gem
Start to bud with verdant sheen.
Healing powers has this plant:
Great Medicine—this we can make:
Tiny leaves, in threes, will grant
Cures for various ills and ache.

But it has another use—
Common copper it will fix,
Silver, cinnabar, produce—
Such the compass of its tricks.

Pan-hsia is *Pinellia tuberifera,* Ten. Its root contains essential oil, fat, and phytosterin besides ash and alkalies. (R711; for the medicinal use of this plant see *PTKM, 17:* 980.2 ff.) Tan-yang 丹陽, originally the name of a city, refers to copper, from a belief that it was here that the alchemist San-Mao-chün 三茅君 once transmuted copper into gold for the relief of the poor.

XVIII. *Chin-ssu lung-ya* 金絲龍芽
(The "Golden Thread Dragon Sprout")
t'u-ssu 兎絲
Growing in the wilderness
It has faintly yellow stems;
Only alchemists can guess
That this plant brings many gems.
Chief among these is the fact
That it works for mercury
With this substance to react:
Sages know this recipe.
Through the wild this plant will sprawl,
Over landscapes primitive,
But will flourish best of all
Where the great Immortals live.

T'u-ssu is *cuscuta japonica,* or dodder, as already established under Verse XIII.

XIX. *Lu-jung lung-ya* 鹿茸龍芽
(The "Deer Antler Dragon Sprout")
lan-ts'ao 藍草

Here and there this may be found,
But the place it favors most,
Where this species will abound,

Is where dry earth acts as host.
Round the year its leaves will thrive,
Flourishing through heat and cold;
Alchemists from this derive
Secrets coveted as gold.
Forming a precipitate,
This, when used by sages true,
Fixes cinnabar in state
And arsenious oxide too.

Lan-ts'ao is probably identifiable with *lan* 藍, *Polygonum tinctorium,* Lour. or indigo plant, the leaves of which contain indigo, kaempferol, kaempferin, acetic acid, and indican (R579). It is included in the *Shen-nung pen-ts'ao ching* as a plant that arrests the process of aging, preventing hairs from turning gray and increasing agility by keeping weight down.

XX. *Yü-p'ing lung-ya* 玉瓶龍芽
(The "Jade Vase Dragon Sprout")
lo-po 蘿蔔

Like a vase of jade this tree
Has a root with section round:
One especial mark we see—
Large the head with which it's crowned.
Nature takes its course when this
Comes in touch with cinnabar:
Sages with this artifice
Cause some strange phenomena.
Handled by the wisest men
Cinnabar will take its cue:
Place in it the root's hole—then
Wonderful results ensue.

Lo-po is *Raphanus sativus,* L., or radish, the seeds of which contain fatty oil, ash, essential oil, sulfuric acid, erucic acid, and $C_9H_{15}NS_2$, while the root contains raphanol, rettichol, essential oil, methylmercaptan, B vitamins, and vitamin C_2,

$C_{10}H_{11}NS$, sinapine, and oxydase. (R482) "*Lo-po*" is the popular synonym for *lai-fu* 萊菔 (*PTKM, 26:* 1192.1 ff.).

XXI. *Yü-ying lung-ya* 玉英龍芽
(The "Jadelike Elegant Dragon Sprout")
liu-hsu 柳絮

Strolling on the river's strand
In the final months of spring,
We can see this wonderland—
Catkins in the breezes swing
Near pavilions bright and neat.
If correctly we collect
And with right procedure treat,
Wonders will the plant effect.
Those who master well this stuff
Hold the heavens in their ken;
Then will fortune bless enough
And Immortals praise such men.

Liu-hsü is catkins from the common willow, *Salix babylonica*, L. (R624; for its medicinal use, see *PTKM, 35B:* 1412.2 ff.).

XXII. *Hsüan-tou lung-ya* 懸豆龍芽
(The "Hanging Beans Dragon Sprout")
tsao-chiao 皂角

Like a knife, or like a sword,
This from high above depends.
Sages know it will afford
Powers that can serve our ends.
Sal ammoniac it subdues,
Fire it can well resist;
But the sage can also use
Other powers on its list,
For when previous tasks are done
Then another attribute
By this dragon sprout is won—
Common copper to transmute.

Tsao-chiao means the pods of *Gleditschia sinensis* or the soapbean tree, the bark of which contains saponin, $C_{59}H_{100}O_{20}$, and arabinose (R387; for its medicinal use see *PTKM, 35B*: 1403.2 ff.). This is included as an edible plant in *CHPT, 2A*: 39a.

XXIII. *Chin-ching lung-ya* 金精龍芽
(The "Golden Essence Dragon Sprout")
ta-chi 大戟

This has fresh and golden leaves:
Lean they look when first they sprout;
And its roots, the sage perceives,
Has within the soil held out,
Growing there for countless years.
In the second month of spring
Vital principle uprears,
Swiftly rising to full swing.
Reaped at this auspicious time
This can fix bright mercury,
Stable will this stuff sublime
In reaction-vessel be.

Ta-chi is *Euphorbia pekinensis*, Rupr. or Peking spurge (R327; for its medicinal use see *PTKM, 17A*: 949.1 ff.).

XXIV. *Wu-yeh lung-ya* 五葉龍芽
(The "Five Leaves Dragon Sprout")
ma-ch'ih 馬齒

Red the bark, and black the seed,
Green the leaves as in the spring,
Five Directions have decreed
Shapes and colors of this thing.
And the stems are white inside,
Yellow flowers grace the plant.
When to cinnabar applied,
Cinnabar it will enchant,
Fix, and change to yellow hue.
Feed this to the cinnabar-field:

Strangest virtue will ensue—
Life eternal it will yield.

Ma-ch'ih is *ma-chih-hsien* 馬齒莧, *Portulaca oleracea*, L. or purslane. It contains vitamin C_1, ash, fat, and urea (R554). Mercury has been obtained from the plant by careful pounding, drying, and autolysis (see Needham, III, 679). It has been also known by the synonyms *wu-fang-ts'ao* 五方草 (five directions plant) and *wu-hsing-ts'ao* 五行草 (Five Elements plant). The former concept is reflected by the five colors mentioned in the text; red can be taken to denote south, black north, green east, white west, and yellow the center. Li Shih-chen says that the alchemists use this plant to fix arsenic, to form mercury, to heat with cinnabar or to fix sulfur, to "kill" realgar and fix orpiment, each according to its own method (*PTKM, 27:* 1212.1 ff.). This plant is also mentioned in *CHPT, 2B:* 31a.

XXV. *Shui-fou lung-ya* 水浮龍芽
(The "Floating-on-Water Dragon Sprout")
fou-p'ing 浮萍

Shaped like coins, the leaves are green,
Water-grown, unhemmed by roots,
Fifth moon is the time serene
For the gathering of these shoots.
Stored and dried they will become
Elements in alchemy;
Patient men may pluck this plum
If they learn the strategy.
Fixing sal ammoniac
Is a task it carries well:
This is its especial knack,
One in which it does excel.

Fou-p'ing is *Lemna minor*, L. or duckweed, which contains iodine and bromine (R702). Its value as a medicine of longevity that reduces the body weight is mentioned in the *Shen-nung pen-ts'ao* pharmacopoeia (*PTKM, 19:* 1068.2 ff.).

XXVI. *T'ung-ting lung-ya* 通頂龍芽
(The "Balding Dragon Sprout")
ku-ching-ts'ao 谷精草

This resembles fallen snow:
Seldom on a barren land
Will this tender flower grow,
But luxurious and grand,
Like the lotus in the pond,
It will thrive where water lies
And to nourishment respond.
Rare the value and the prize
To the gifted Alchemist
Who can recognize its worth:
Cinnabar, with just a twist,
Turns goose-colored in rebirth.

Ku-ching-ts'ao is probably an abbreviated form for *ku-ching-ts'ao* 穀精草, *Eriocaulon australe*, R. or pipewort (R701). The *Ta Ming* pharmacopoeia says that it combines with mercury to form cinnabar (*PTKM, 16:* 936.2 ff.).

XXVII. *Ti-ku lung-ya* 地骨龍芽
(The "Earth Skeleton Dragon Sprout")
chu-ch'i 枸杞

Fruits resemble cinnabar,
Cloudlike, as they deck this vine;
Also cloudlike, nebular,
Are the leaves that intertwine.
Winter-nurtured are the roots,
But in spring and summer days
Leaves and flowers are the fruits.
In the autumn season's haze
Maidens red we may collect.
When with cinnabar we churn,
This will have a strange effect—
Silver color it will turn.

Chu-ch'i is *Lycium chinense,* Mill. or matrimony vine. Its leaves contain betaine and choline and its stem HCN and ash (R115). The *Shen-nung pen-ts'ao* describes it as a medicine of longevity. Stories are told about the efficacy of its roots, which after a long time are said to assume the shape of a dog (*PTKM, 36:* 1452.1 ff.). The Yuan pharmacopoeia *T'ang-i pen-ts'ao* 湯液本草 says that it fixes sulfur and cinnabar (*PTKM, 36:* 1453.1). "Maidens red" must refer to the fruits. See also *CHPT, 2A:* 37a.

XXVIII. *Ti-ting lung-ya* 地頂龍芽
(The "Land Attendant Dragon Sprout")
ch'e-ch'ien-tzu 車前子

Land Attendant bears no roots
Though it bears some tender leaves.
Single stem from earth upshoots
This the housewife well receives.
Fixing mercury, and then
Turning it into "the most
Precious substance" known to men—
Of this power it can boast.
How can common men attain
Magic skills this plant can give?
How can they its secrets gain
And like great Immortals live?

Ch'e-ch'ien-tzu, Plantago major, L., or greater plantain, was called "cart-front" because of its occurrence along the wayside near the dung left by horses and cattle. Both the plant and its seeds are used for medicinal purposes. The whole plant contains choline, potassium salts, adenine, citric acid, oxalic acid, vitamin C, and aucubin, while its seeds contain resin and plantagin (see R90). The *Shen-nung pen-ts'ao* regards it as a medicine of longevity. T'ao Hung-ching says that it lightens one's body so that one can jump across valleys and that it confers immortality (see *PTKM, 16:* 918.2 ff. and *CHPT, 1A:* 7a).

XXIX. *Mu-erh lung-ya* 木耳龍芽
(The "Woody-Ear Dragon Sprout")
fo-erh-ts'ao 佛耳草

Looking like a human ear,
This strange plant will grow on wood:
Though the Sage will this revere
Little it is understood.
Minerals Eight the plant will change,
Medicine Great will then result.
Other powers in its range
Make the sage's heart exult.
Just a little of the stuff,
Add this to realgar true,
This is magic strength enough—
Gold appearance will ensue.

The term *fo-erh-ts'ao* was first used as a synonym for *Shu-ch'ü-ts'ao* 鼠麴草, *Gnaphalium multiceps*, Wall., in the sixteenth-century pharmacopoeia *Pen-ts'ao p'in-hui ching-yao* (*13:* 401). However, the context of the stanza itself seems almost certainly to refer to *mu-erh* 木耳 (literally, woody ear), or *Auricularia Auriculajudae*, Schr. (R827A).

XXX. *Ti-p'an lung-ya* 地盤龍芽
(The "Earth-Basin Dragon Sprout")
ho-yeh 荷葉

By projection, calomel
Changes into silver bright.
Shining green and rounded well,
Leaves present a pretty sight,
Float on surface of a lake.
Often seen by common men—
Nothing much of this they make—
This, within the scholar's den,
Magic powers will confer.
Fixing arsenic with this herb,
Good results will soon occur,
Bringing side effects superb.

Ho-yeh refers to the leaves of the *Nelumbo nucifera*, Gaertn. or lotus leaves. It contains nelumbine (R542). Its medicinal use is described in *PTKM*, *33:* 1342.1 ff.

XXXI. *Wan-chang lung-ya* 萬丈龍芽
(The "Hundred-Thousand-Feet-Long Dragon Sprout")
sung-lo 松蘿

Hanging from the tall pine tree,
These display majestic strength,
Several mossy strands we see,
Hundred thousand feet in length.
Here the flowers may be few,
There in clusters they will grow;
Cinnabar is made anew—
Great Medicine from this will flow.
Other deeds this can perform:
Fed with this—without a word—
Cocks and crows will take the form
Of the noble phoenix bird.

Sung-lo is pine lichen, *Usnea diffracta,* Vain. It contains usnic acid, $C_{18}H_{16}O_7$, barbatic acid, $C_{19}H_{20}O_7$, methylbarbatic acid, and ranalinic acid (R821). This parasite has been confused with *t'u-ssu-[tzu]*, *Cuscuta japonica*, Choisy (*PTKM, 37: 1475.1 ff.*).

XXXII. *Tzu-hua lung-ya* 紫華龍芽
(The "Purple-Flower Dragon Sprout")
tz'u-chi 刺薊

Purple-Flower Dragon Sprout
Grows in long-deserted farms;
Stems are green in hue throughout—
It performs some magic charms.
Gathered are the leaves and roots,
Though the plant be young or old,
And mid-summer sun most suits—
Drying will be well controlled.
Arsenic this can fix with ease;
Mercury will submit as well;
And the plant, in fixing these,
Holds them in a constant spell.

Tz'u-chi is a synonym of *ta-chi* 大薊, *Cnicus spicatus*, Max., or the tiger thistle (R30). See *PTKM, 15:* 866.1, and *CHPT, 1A:* 1b.

XXXIII. *Hsiang-mu lung-ya* 香木龍芽
(The "Fragrant-Wood Dragon Sprout")
ch'un-mu 椿木

Evergreen its branches spread
In the courtyards and the parks;
When the summer comes, it's said,
Then the Alchemist embarks
On collecting leaves and roots.
Fixing arsenic—strange enough—
This, in doing, it transmutes
To a snow-white powdery stuff.
Other tasks the plant performs,
Unexpected and occult:
Copper, iron, it transforms—
Rarer substances result.

Ch'un-mu is *Cedrela sinensis*, Juss., the fragrant cedar. Its leaves contain protein and fat (R334), but its wood is inferior as a timber (*PTKM, 35A:* 1388.2).

XXXIV. *Chin-jui lung-ya* 金蘂龍芽
(The "Golden Stamens Dragon Sprout")
chü-hua 菊華

Most unusual are the stems
And the roots and flowers too,
Late in autumn will these gems
Show their beauty shining through,
When corolla opens wide.
Sulfur this can fix with ease—
Sulfur will be modified
By the strangest strategies:
Precious substance will appear.
Better than the doctor's pills,

This, with ease, will swiftly clear
Common cold and all its ills.

Chü-hua is *Chrysanthemum sinense*, Sab., or chrysanthemum blossom, which contains essential oil, adenine, choline, and stachydrine (R27). Regarded as a medicine of longevity in the *Shen-nung pen-ts'ao*, one of its uses is given by Ko Hung citing Liu Sheng's 劉生 procedure for elixir making (Ware, p. 88; *PTKM, 15:* 845.1 ff.; *CHPT, 1B:* 75a.)

XXXV. *Wu-tou lung-ya* 烏荳龍芽
(The "Black Bean Dragon Sprout")
hei-tou 黑荳

High in far-off mountain range
Grows this little evergreen
In a landscape somewhat strange,
By the common man unseen.
Only men of science can know
Of the value of this plant—
What its branches, leaves bestow
And the powers they will grant.
Cinnabar is fixed, and so
Shining luster it attains,
But the plant no change will show—
Same in color it remains.

Hei-tou is *Glycine soja*, S. *et* Z., black soybean, commonly called *ta-tou* 大豆 (R388 and *PTKM, 24:* 1134.1). It was believed that by taking these beans as a drug one could go without food for many days (see for example *PTKM, 24:* 1135.1).

XXXVI. *Yuan-yeh lung-ya* 圓葉龍芽
(The "Round Leaf Dragon Sprout")
hsien-ling-p'i 仙靈脾

Mountains blue, of giddy height—
Here this plant has its estate,
And presents a curious sight:

Little leaves conglomerate.
Summertime is best to choose
To collect and then to dry;
Leaves which alchemists can use,
And their secret art apply.
Fixing red realgar turns
This disulfide into gold;
Through nine cycles mixture churns—
Great Elixir will unfold.

Hsien-ling-p'i is a synonym of *yin-yang-huo* 淫羊藿, *Epimedium macranthum,* Morr. *et* Dene. (R521 and *PTKM, 12B:* 750.2 ff.). It is included as an edible plant in *CHPT, 1A:* 28b.

XXXVII. *Hsiang-fu lung-ya* 香附龍芽
(The "Fragrant-Aconite Dragon Sprout")
so-ts'ao 莎草

Roots are called "Green Aconite."
Long their fragrance will remain;
Near the flowers, in delight,
By the brooks, their choice terrain,
They abound in joyful swarm.
Arsenic they can fix in state
And can also change its form—
Liquid will precipitate.
This, imbibed, confers strange grace,
Making life as long as wide
Rivers flowing in the place
Where Immortals do abide.

So-ts'ao is *Cyperus rotundus,* L. or nutgrass, the roots of which contain essential oil, cyperen, and cyperol (R724). For its medicinal use see *PTKM, 14:* 822.2 ff.

XXXVIII. *Tz'u-sha lung-ya* 慈砂龍芽
(The "Magnetic Sand Dragon Sprout")
t'ien-nan-hsing 天南星

Strange the inflorescence here—
Scallion flowers are akin,
But ere spadix will appear,
Color red will soon begin.
Gathered are the stems and roots—
These have magic properties:
Fixing red realgar suits,
Medicine Great is made with ease.
Further powers has this plant—
Copper base and iron dead
Will this magic herb enchant,
Precious stuff will come instead.

T'ien-nan-hsing is *Arisaema thunbergii,* Bl. or Jack-in-the-pulpit, and is known as *hu-chang* 虎掌, "tiger's paw," in the *Shen-nung pen-ts'ao* (R709 and *PTKM, 17B:* 977.1 ff.). It was first called *t'ien-nan-hsing* in Ma Chih's *K'ai-pao pen-ts'ao,* written in the second half of the tenth century. See also *CHPT, 1B:* 24b.

XXXIX. *I-hua lung-ya* 異華龍芽
(The "Strange Flower Dragon Sprout")
shao-yao 芍藥

At the "grain rains," late in spring,
This will flower to the full,
With each fortnight witnessing
Blossoms that are fanciful.
Flowers, leaves, and roots are all
Gathered and transported home;
Common calomel will fall
Under its prevailing foam.
From a powder it will turn
To a liquid silver-bright.
Do not fear, have no concern,
That this process is not right.

Shao-yao is *Paeonia albiflora,* Pall., the Chinese peony. Its roots contain asparagine and benzoic acid (R536). T'ao Hung-ching mentioned that it was ingested by Taoists during his time (*CLPT, 8:* 201.2, and *PTKM, 14:* 802.2 ff.). The "grain rains" fall on about April 20 each year.

XL. *Lung-pao lung-ya* 龍寶龍芽
(The "Dragon-Treasure Dragon Sprout")
mu-tan 牡丹

Also blooming each fortnight
At the "grain rains' " happy time,
This presents a pretty sight—
Red the buds in ripened prime.
Of the flowers in the field
This is given special place:
Fleets of thousand ships will yield
To the beauty of its face.
Into powder this is made,
Many colors will it show,
Alchemists can use this aid
And immortal life may know.

Mu-tan is *Paeonia moutan,* Sims., the tree peony. Its roots and bark contain paconol, $C_9H_{10}O_3$, essential oil, saccharose, galactose, arabinose, glutamine, arginine, paeonia-fluorescin, glucoside, and enzymes (R537). Wu P'u 吳普 says that prolonged ingestion of the tree peony will decrease the body weight but lengthen the life span (*PTKM, 14:* 804.2 ff.).

XLI. *Yung-ch'ing lung-ya* 永青龍芽
(The "Evergreen Dragon Sprout")
sung 松

Up the mountains blue this grows,
High the cliffs where it will haunt;
Green the leaves, they proudly pose,
Shape of phoenix tail they flaunt.
Quicksilver to fix in state
By projection, this we see;
Humble metals are made great
By its secret alchemy,

Through its grace a thousandfold
Alchemists with ease amass
Timelessness desired of old,
Though a myriad years may pass.

Sung is the general name for pine trees, *Pinus*. The seeds of
Pinus thunbergii, Parl., the black wood pine, contain arginine;
the leaves of *Pinus densiflora,* the red wood pine, contain essential
oil, bornyl acetate, and glucokinin. The branch of the former
contains camphene and the bark of the latter contains glucokinin. The branches of both contain turpentine, *d-x*-pinene,
e-x-pinene, camphene, and sesquiterpene. The exudate of the
black wood pine contains essential oil and resin, while that of
the red wood pine has in addition bornyl acetate (R789). The
Shen-nung pen-ts'ao regards the seeds as a food of immortality
(*PTKM, 34:* 1351.1 ff.).

XLII. *Ts'e-pai lung-ya* 側栢龍芽
(The "Inclined Arbor Vitae Dragon Sprout")
pai 栢

In the far-off mountain shades
And along the sparkling stream,
Filling up the happy glades—
Here ten thousand clusters teem.
Great the fame that this enjoys:
Of the known ten thousand plants
This is what the sage employs.
Wondrous are the gifts it grants:
Fruit confers longevity;
Arsenic it can fix outright,
Turning it, with subtlety,
To a silver that is white.

Pai is *Thuja orientalis,* L., or arbor vitae. Its kernel contains
fat, its leaves pinene and caryophyllene, and its exudate
essential oil and resin (R791). Again it is regarded as a food
of immortality by the *Shen-nung pen-ts'ao* (*PTKM, 34:* 1349.1
ff., and *CHPT, 2A:* 38a).

XLIII. *Chin-hua lung-ya* 金華龍芽
(The "Golden Flower Dragon Sprout")
k'uei 葵

In the genial months of spring
Grows this truly lovely plant;
Soon, with growth and ripening,
Yellow flowers will enchant.
When this Medicine Great succeeds
Magic power knows no bounds;
It can satisfy our needs—
This the Alchemist propounds.
Common people cannot know
How to find this secret key;
Search where great Immortals go—
There you'll find the recipe.

K'uei is *Malva verticillata*, L., Chinese mallow (R280). The *Shen-nung pen-ts'ao* says that prolonged administration of this drug strengthens the bones, grows muscles, decreases the body weight, and increases the life span, and T'ao Hung-ching recommends it as an antidote to elixir poisoning (*PTKM, 16: 902.2 ff.*).

XLIV. *Tzu-chin lung-ya* 紫金龍芽
(The "Purple Gold Dragon Sprout")
chang-liu 章柳

Purple roots, and white the leaves—
This can grow beside the tree.
Knowing Alchemist receives
From forbears this recipe.
Minerals, metals, are the theme:
Fix them and react with them—
These will fall within the scheme;
And, to crown this stratagem,
They evaporate away.
Add to sulfur just a bit,
Sulfur bows beneath its sway,
To its power must submit.

Chang-liu is *Phytolacca acinosa,* Roxb. or poke root. It is known as *shang-lu* 商陸 in the *Shen-nung pen-ts'ao.* Its roots contain phytolaccatoxin and saponins (R555). Su Ching says that it was much used by the alchemists of old, while the *Ta Ming pen-ts'ao* reports that it can fix sal ammoniac, arsenic, and orpiment and also attract lead (see *PTKM, 17A:* 944.2 ff.).

XLV. *Ti-tan lung-ya* 地膽龍芽
(The "Earth-Gall Dragon Sprout")
t'u-ssu-tzu 兔絲子

Purple flowers adorn this tree,
And its shapely leaves are green;
Strange the plant's ecology:
Farms are where it may be seen.
Arsenic fixing is its forte;
Sulfur it will enter through,

[The original stanza is incomplete and abruptly ends here. We continue the stanza in the spirit of a Chinese alchemist disappointed over the missing lines.]

(In the Alchemist's retort:
Magic lies within this brew.
Little else we know of this:
Does it bring the sages gold?
Give to them eternal bliss?
Answers never will be told.)

T'u-ssu tzu is *Cuscuta japonica,* which we have already come across in verses XIII and XVIII.

XLVI. *T'ieh-so lung-ya* 帖索龍芽
(The "Dangling Rope Dragon Sprout")
yang-chiao-miao 羊角苗

[The original stanza is completely missing here. We compose the following on behalf of a Chinese alchemist in dismay.]

(Like the previous plant discussed,
Mystery shrouds this Dragon Sprout;

Hidden deep within the dust,
Lies the truth beyond a doubt.
But we cannot hope to know
More than nature cares to tell.
Where and how does this plant grow?
What its shape, and what its smell?
These are secrets lost to men:
"Dangling Rope" is what it's called:
Secrets stay in scholar's den,
Never to be unenthralled.)

Yang-chiao-miao is not described in the pharmacopoeias but is mentioned in the *Chiu huang pen-ts'ao* (*1B:* 19a) and the *Yeh-ts'ai po lu* 野菜博錄 (*2:* 9b). It is described as a wild vine with green stems and white leaves, which when fully grown turn green on their upper surfaces but remain white in the lower. Its leaves grow in opposed pairs, and when they are torn white sap flows out. Stalks emerge from between the leaves bearing small white five-petaled flowers resembling the horn of a goat. Its leaves have a sweet but slightly bitter taste. Pao Shan says that it looks like *shan yao* 山藥, which is Chinese yam (R657); hence it probably belongs to the *Dioscorea* family.

XLVII. *Ching-t'u lung-ya* 淨土龍芽
(The "Clean Earth Dragon Sprout")
tu-chou-ts'ao 獨箒草

Quite uncultivated, this
Grows in households, and will line
Long pavilion's edifice:
Wild it is but yet benign.
It communes with sun and moon
And in doing this it turns
Great Medicine—to men a boon;
This is what the scholar learns.
Other powers has this plant;
From its gifts the gains are great:
Substances it can enchant,
Cinnabar to fix in state.

Tu-chou-ts'ao is a synonym of *ti-fu* 地膚, *Kochia scoparia,* Schrad., the broom plant, the leaves of which contain saponin (R562). It is regarded as a medicine of longevity in the *Shen-nung pen-ts'ao.* Li Shih-chen says that when the shoot and the leaves are reduced to ash by fire, it fixes arsenic, calomel, mercury, sulfur, realgar, and sal ammoniac (*PTKM, 16:* 913.2 to 914.2; *CHPT, 1A:* 33b.

XLVIII. *Tao-sheng lung-ya* 道生龍芽
(The "Wayside-Born Dragon Sprout")
ti-pien-chu 地編竹

Small the leaves, the flowers red,
By the wayside it will grow;
That Immortals eat this bread
Passers-by need never know.
With white arsenic to react—
This is quite within its skill:
From the force of this compact,
Liquid will the plant distill.
Just a little of the stuff
On base copper renders it
Soft and smooth as woolly fluff:
Copper must to plant submit.

Ti-pien-chu is probably a synonym of *p'ien-hsu* 篇蓄, *Polygonum aviculare,* L., knotweed or goose-weed. This plant contains tannin, resin, wax, essential oil, flavone, sugar, and zinc besides ash (R566). Li Shih-chen reports that it was used by the alchemists, first burned and reduced to ash and then made into a sublimate (*PTKM, 16:* 934.1,2 and *CHPT, 1A:*4a).

XLIX. *Hsien-i lung-ya* 仙衣龍芽
(The "Immortal Dress Dragon Sprout")
sung-lo 松蘿

Hanging down from awful height
Very lengthy is this vine;

This is Alchemist's delight
And a plant of rare design.
Gathered in the mountain range,
This has many subtle arts;
States of matter it can change:
Other features it imparts.
Fixing metal, minerals—these
Turn to airy vapor's breath;
Mercury it can fix with ease,
Sulfur it can put to death.

Sung-lo is *Usnea diffracta,* Vain., or pine lichen. The plant contains usnic acid, $C_{18}H_{16}O_7$; barbatic acid, $C_{19}H_{20}O_7$; methylbarbatic acid; and ramaline acid (R821; *Journal of Japanese Botany*, 1965, *40:* 173; ibid., 1967, *42:* 289). For the medicinal use of this parasite see *CLPT, 9:* 230.2, and *PTKM, 37:* 1475.1; see also stanza **XXXI**.

L. *Ch'ih-chao lung-ya* 赤爪龍芽
(The "Red Claw Dragon Sprout")
po-ts'ai 波棶

Great the heat this can withstand,
Frost intense it can resist;
Secret powers come to hand
For the learned Alchemist.
But beware when strong winds blow,
For to these the plant gives way
And is injured and brought low,
Falling into disarray.
Arsenic fixing is its art—
Jadelike substance will result.
Dear to great Immortal's heart,
This is used in ways occult.

Po-ts'ai is *Spinacia oleracea,* L., spinach. It contains saponin, protein, fat, carbohydrate, ash, iron, calcium, iodine, copper, magnesium, spinacin, vitamins A_1, B_1, C_3, and G_2, urea, chlorophyll, chlorophyllase, hentriacontan, $C_{31}H_{64}$, cerotinic acid, linolenic acid, malic acid, citric acid, lecithin, sterines, and betaine (R563). In the Sung pharmacopoeia *Chia-yu pen-ts'ao* it is called *po-ling* 菠薐, and its synonym *po-ts'ai* appears in the *Pen-ts'ao kang mu* (*27:* 1207.1). Meng Shen says that it is beneficial to those who consume elixirs.

LI. *Hsien-chang lung-ya* 仙掌龍芽
(The "Immortal's Palm Dragon Sprout")
ts'ang-p'eng 蒼蓬, *chin-kou* 金勾, *suan-tsao* 酸棗

Growing near the frigid pool
And the cave where hermit prays,
This is gathered as a rule
In the early summer days
When it's fully ripe and rich.
Made into a sublimate,
To a powder it will switch:
Sulfur it will fix in state.
But this art is little known—
When it's eaten change occurs:
Hair will turn to darker tone,
Youthful looks the plant confers.

Three names are listed together under this item on the "Immortal's Palm Dragon Sprout," namely *ts'ang-p'eng, chin-kou,* and *suan-tsao.* Presumably all three must mean the same plant. Nothing is found regarding the first two names, but *suan-tsao* is the wild jujube, *Zizyphus vulgaris,* Lam., var. *spinosus,* Bge. (R294). The *Shen-nung pen-ts'ao* regards its fruit as reducing body weight and lengthening human life (*PTKM, 36:* 1440.2). *Suan-tsao* is mentioned in *CHPT, 2A:* 24b.

LII. *Yin-fa lung-ya* 銀髮龍芽
(The "Silvery Hair Dragon Sprout")
ts'ung 葱

Like the garlic roots ornate,

So the roots of this strange tree
Human hair approximate:
Such its rare morphology.
Arsenic it will fix and stay
From subliming, and it's true
Common folk in every way
Fail to see what it can do.
Medicine of immortals great,
This has other gifts as well:
Arsenic fixing, correlate—
End result is calomel.

Ts'ung is *Allium fistulosum*, L., the Chinese small onion, or ciboule. It contains vitamins B_1, C_2, and G_1, protein, fat, carbohydrate, ash, and small amounts of copper and magnesium (R666). T'ao Hung-ching says that the juice was used by the alchemists for making aqueous solutions or suspensions of jade and various minerals (*PTKM, 26:* 1175.1 ff., esp. 1177.2).

LIII. *Hsien-li lung-ya* 仙力龍芽
(The "Immortal Power Dragon Sprout")
chiu 韭

Copper fixing is its art,
Silvery stuff its end result;
Men who seek immortal part
In this substance will exult.
Mention not this mystic plant
To the earthlings whom you know,
Make no man your confidant
Though he may be friend or foe.
Gathered are the potent leaves
When they ripen in the sun,
Ills of age the plant relieves
When its magic work is done.

Chiu is *Allium odorum*, L., leek. It contains protein, fat, carbohydrate, ash, and vitamin C_2 (R670). For its medicinal use see *PTKM, 26:* 1172.1 ff.

LIV. *Ch'ang-sheng lung-ya* 長生龍芽
(The "Longevity Dragon Sprout")
nai-tung 奈凍

Though its bloom is yet to show,
This outshines the flowers around;
Tender sprouts will gaily blow,
By the pools they may be found.
Sulfur it will fix and change
Into stuff that looks like jade;
Nature's ways to rearrange—
For this task we use this aid.
Other arts within its ken:
Copper, iron to refine
To a stuff to please all men—
Noble metal rare and fine.

Nai-tung is probably identifiable with its homophone, and indeed in a way its homonym, *nai-tung* 耐冬, which itself is a synonym of *lo-shih* 絡石, *Trachelospermum divaricatum*, K. Sch. (R168). It is regarded by the *Shen-nung pen-ts'ao* as a medicine of longevity (*PTKM, 18B:* 1049.2).

LV. *San-huang lung-ya* 三黃龍芽
(The "Three Yellow [Substances] Dragon Sprout")
ti-huang 地黃

On uncultivated land,
These will grow beside the brook;
Learn to read the secret hand,
Through this mystery learn to look—
Yellow Substances, all three,
Teach us to communicate
With the spirits that are free.
Fixing sulfur in its state,
Essence it will keep intact.
But if medicine is to work
Fire on the plant must act;
Secrets which in this plant lurk
Only fire can extract.

Ti-huang is *Rehmannia glutinosa,* Lib. (R107). The *Shen-nung pen-ts'ao* regards it as a medicine of longevity (*PTKM, 16:* 892.1 ff.). Ko Hung says that the adept Ch'u Wen-tzu 楚文子 after taking this substance for eight years could see things shining in the night (see Ware, p. 196, under Prince Wen of Ch'u).

LVI. *Ch'an-shu lung-ya* 纏樹龍芽
(The "Tree-Entwining Dragon Sprout")
ling-hsiao 凌霄

Seeking the empyrean height
This has roots that perch and cling;
Circling trees and holding tight,
Spreading out and posturing.
Summer is its time of boon
When its pretty blooms are seen.
Gathered in the seventh moon,
Precious medicine we may glean.
But the plant has other arts:
Mercury's essence it can turn
On projection—dross departs—
Whitish silver we discern.

Ling-hsiao is *Tecoma grandiflora,* Loisel (R101). It is known as *tzu-wei* 紫葳 in the *Shen-nung pen-ts'ao* (*PTKM, 18A:* 1016.1).

LVII. *She-sheng lung-ya* 舍生龍芽
(The "House-Grown Dragon Sprout")
wa-sung 瓦松

Not on earth is this plant found
Nor among the mountains high;
On the houses they abound,
And in crevices they lie,
On the roof tops interfused.
Gathered by the Alchemist,
In reaction vessel used,
Fierce the fires they can resist.
On to mercury we place

Just a little of this fern:
Mercury will dry apace—
Into silver it will turn.

This plant has not yet been properly identified. The *Chung yao chih* 中藥志 does not specify a variety of the species. Read gives it as *Cotyledon fimbriata*, Turcz., *var. ramosissima*, Maxim., which is identified by Steward as yet another plant, *hsu-shih-lien-hua* 鬚石蓮花 (see R469 and A. N. Steward, *Manual of Vascular Plants of the Lower Yangtse Valley of China*, Japan, 1958). Smith also refers to it as *Cotyledon fimbriata*, saying that it is used as a styptic for dysentery, as an ointment for falling out of the eyebrows, as a stimulant for suppressed menstruation, for gravel, and for dog bite (see F. P. Smith, *Chinese Materia Medica*, revised by G. A. Stuart, first published Shanghai, 1911, reprinted Taipei, 1969, p. 449). It was first listed in the T'ang phamacopoeia under the name *tso-yeh ho-ts'ao* 昨葉何草 (*PTKM, 21:* 1089.1).

LVIII. *Nai-tung lung-ya* 耐凍龍芽
(The "Cold-Enduring Dragon Sprout")
hsia-k'u-ts'ao 夏枯草

Dense the leaves which deck this plant,
Into sections they divide;
Pale and purple flowers enchant,
In the springtime is their pride,
Though in summer they will fade.
Gathered by the men who know
This becomes a wonder aid—
Many gifts it will bestow,
Juice extracted is the key:
Mercury's essence it will char—
From this impact we shall see
Desiccated cinnabar.

Hsia-k'u-ts'ao is *Prunella vulgaris*, L., the heal-all or carpenter weed. Bitter principle and essential oil are obtained from the herb (R. N. Chopra et al., *Glossary of Indian Medicinal Plants*, New Delhi, 1956, p. 204, and R122; for its medicinal use see *PTKM, 15:* 860.1). The *Shen-nung pen-ts'ao* says that it decreases the body weight. See also *CHPT, 1A:* 18b.

LIX. *Sang-sun lung-ya* 桑笋龍芽
(The "Mulberry-Cone Dragon Sprout")
sang yeh 桑葉

This is cut in early spring
And to ash is fully burned;
Into water this we fling
And, dissolving, it is churned.
Thus it will recrystallize
And will turn into a base.
Heat with sulfur, energize,
Scarlet flame will show its face.
If we add a little bit
To the mercury in tray,
Mercury must needs submit,
And will not distill away.

Sang sun and *sang yeh* refer to the fruits and the leaves of the mulberry tree *Morus alba*, L. respectively. The former contain fat and urease, while the latter contain calcium malate, calcium carbonate, invert sugar, pentosan, tannin, carotene, vitamin C_2, and cholin (R605). Su Ching reports that the leaves were ingested by the aspirants to immortality, and, collected during the tenth lunar month in early winter, were called *shen-hsien yeh* 神仙葉, "leaves of the immortals" (*PTKM, 36:* 1429.1 ff.). See also stanza II.

LX. *Chung-yang lung-ya* 中央龍芽
(The "Central Dragon Sprout")
huang-ts'ao 黃草

Treasured as a rarity
In the villages afar,
This is boon to alchemy,

Hides a magic formula.
Into powder it is ground
And to ash we must reduce;
Cauterized and thus embrowned
It is rendered fit for use.
If we add to mercury
Strange effects it will create:
Mercury will boil with glee
But will not evaporate.

Huang-ts'ao is a synonym for *ai* 艾, which is *Artemisia vulgaris*, L.,
the common mugwort, but it can also be a synonym for *chin
ts'ao* 藎草, *Arthraxon ciliaris*, Beauv. The leaves of the former
contain essential oil with cineole (R9 and R732). The former
has been used for cauterization and the latter as a dye. (For
their medicinal uses see *PTKM*, *15:* 848.2 ff. and *16:* 934.2
ff.) It is not possible to say with any certainty which of the two
the text refers to.

LXI. *Hsuan-ch'iu lung-ya*玄毬龍芽
(The "Mysterious Ball Dragon Sprout")
ch'ieh-tzu 茄子

When in season fruits appear
In the flowers' company,
Each is fashioned like a sphere:
Hence the name of this strange tree.
Fixing sulfur is its gift;
It can change base copper too
And from lowly state will lift—
Snow-white substance will ensue,
To those whom such arts concern
Everlasting life will reach;
Common earthlings cannot learn
What the sages have to teach.

Ch'ieh-tzu is the eggplant or brinjal, *Solanum melongena*, L. It
contains trigonelline, choline, and vitamins A_1, B_1, and C_2,

while the fruits contain fat and protein (R119). It first appears in the tenth-century pharmacopoeia *K'ai-pao pen-ts'ao* (*PTKM, 28:* 1230.1 ff.).

LXII. *Hsiang-lu lung-ya* 香爐龍芽
(The "Incense-Burner Dragon Sprout")
tzu-su 紫蘇

Yellow flowers, verdant leaves,
Grow in artificial ponds;
Sulfur in its grip upheaves
And a new appearance dons
Staying fixed by its strange art.
Minerals Eight will lose their power
And their properties depart
Under influence of this flower.
Great Immortals left behind
This procedure for our gain,
But the common people find
Secrets they cannot unchain.

Tzu-su, known earlier by T'ao Hung-ching by the name *su* 蘇, is *Perilla nankinensis,* Decne. Its leaves contain essential oil, perillol, resin, and fatty oil (R135). For its medicinal use see *PTKM, 14:* 840.2 ff. See also *CHPT, 2B:* 53b.

LXIII. *Ch'ing-lung lung-ya* 青龍龍芽
(The "Empyrean Dragon Sprout")
ko-ken-man 葛根蔓

Vines will sprout in early spring,
And a dragon's head will show;
Round and round the vines will ring,
All directions they will grow.
Under its mysterious art
Minerals Eight will change their grade,
Virtue they will then impart:
Medicine Great is thereby made.
Mercury it accompanies:

Mercury will coagulate
And will then react with ease,
Changing to a silvery state.

Most probably *ko-ken-man* refers to the vine of *Pueraria hirsuta,*
Schneid. (R406). The hemp itself is usually called *ko* 葛 and
its bulbous root *ko-ken* 葛根. The latter, because it is edible and
has commercial value, is regarded as the most important part
of the plant, and sometimes in everyday usage the term represents the plant as a whole. Thus *ko-ken-man* would mean its
vine. The pharmacopoeias also mention another vine with a
somewhat similar name, *ko-le-man* 葛勒蔓, a synonym of *lü-ts'ao*
葎草, *Humulus japonicus,* S. *et* Z., wild hops (see R604 and
PTKM, 18B: 1048.2).

LXIV. *Ti-shen lung-ya* 地參龍芽
(The "Earthly Ginseng Dragon Sprout")
chih-mu 知母

Mixed with *Fritillaria*
Verticillata, and some
Other substance we have here—
Arisaema japonicum—
Ground into a powder fine,
Saturate with vinegar,
With some common grass combine,
Wrap this round some cinnabar;
In reaction-vessel place,
Roast the mix in special fire,
Then the stuff will change its face:
Silver will we thus acquire.

Chih-mu is *Anemarrhena asphodeloides,* Bge. (R675). For the
medicinal use of this plant see *PTKM, 12A:* 736.1. *Pei-mu*
貝母 is *Fritillaria verticillata,* Willd., containing peimine and
peiminine (R678). For its medicinal use see *PTKM, 13:* 780.1.
Nan-hsing 南星 refers to *t'ien-nan-hsing* 天南星, already discussed
under XXXVIII. See also *CHPT, 1B:* 25b.

LXV. *Tzu-pei lung-ya* 紫背龍芽
(The "Purple Spine Dragon Sprout")
yu-tien-yeh 油點葉

Humulus Lupulus this is called;
Some have named it "Purple Spine."
Rare these plants that have enthralled
Students of the science divine.
Fruits are small and somewhat lean;
Flowers red and spotted leaves
On this rare plant may be seen.
Through its use the sage achieves
Powers that will stupefy.
Cinnabar by this is slain;
Mercury it will fix and dry,
Silver in this way we gain.

No reference is found about *yu-tien-yeh*. However, the synonym
tzu-pei lung-ya 紫背龍芽 refers to *she-han* 蛇含, *Potentilla kleiniana,*
Wight (R442; *PTKM, 16:* 921.1).

LXVI. *T'ien-yen lung-ya* 天焰龍芽
(The "Celestial Flame Dragon Sprout")
lien-hua 蓮華

Roots and leaves in water grow,
Flowers are like roaring flames,
On the backs of leaves will glow
Colors red of many names.
Under sixth moon is the time
For the gathering of this tree;
Then the plant is in its prime
And is fit for alchemy.
By its use we can create
Matter of unusual strength;
Mercury it will fix in state;
Silver will appear at length.

Lien-hua is the flower of the lotus, *Nelumbo nucifera,* Gaertn.

(R542). T'ao Hung-ching mentions that this was used by aspirants to immortality, especially for the making of incense (*PTKM, 33:* 1341.2).

LXVII. The Song of Preserving the Natural Endowment

When the Dragon Sprouts we heat,
Smoke must issue well controlled;
When the smoke becomes effete,
Then the magic will unfold;
Essence of this mystic tree
Will be greatly fortified.
Difficult the alchemy—
Special fire must be applied,
Five pounds of fuel must be burnt.
Such mysterious regimen
Can with diligence be learnt
But is closed to common men.

LXVIII. On the Identification of the Genuine (Substances)

Only at the proper time,
When its season has begun,
When the sprouts are in their prime,
May the gathering be done.
If we use the genuine stuff
And the sage's rules obey,
We shall have success enough—
Failure will not come our way.
Roots and leaves and fruits and stems
Must be used with due respect.
Wrong procedures, stratagems,
Bring on danger and neglect.

LXIX. For the Predestined

Now we come to journey's end,
And a rest we well deserve;
These the verses we have penned

Must a useful cause subserve.
Dragon Sprouts are Seventy-Two:
Gathered at the proper time,
Treasured well as is their due,
They will bring us gifts sublime:
Noble stuff from base we gain
And this wealth from heaven sent
May to purposes humane
Be applied and wisely spent.

Bibliography

Asahina Yasuhiko 朝比奈泰彦 et al.
Shōsōin yakubutsu 正倉院藥物. Osaka, 1955.

Chu Shu 朱橚
Chiu huang pen-ts'ao 救荒本草 (1406). Reprint of the 1525 edition, Shanghai, 1959.

Ho Ping-yü and Needham, Joseph
"Theories of Categories in Early Mediaeval Chinese Alchemy," *Journal of the Warburg and Courtauld Institutes*, 1959, *22*: 173–210.

K'ung Ch'ing-lai et al.
Chih-wu-hsueh ta tz'u-tien 植物學大辭典. Shanghai, 1918.

Li Shih-chen 李時珍
Pen-ts'ao kang mu 本草綱目 (1596). Reprint from the Wei ku chai ed. of 1885, Peking, 1957.

Liu Wen-t'ai 劉文泰 et al.
Pen-ts'ao p'in hui ching yao 本草品彙精要 (ca. 1505). Shanghai, 1959.

Needham, Joseph
Science and Civilisation in China. Vol. III. Mathematics and the Sciences of the Heavens and the Earth. Cambridge, England, 1959.

Pao Shan 鮑山
Yeh ts'ai po lu 野菜博錄 (1622). Reprint from the first edition in *Ssu pu ts'ung k'an* 四部叢刊 series.

Read, Bernard E. [and Liu Ju-ch'iang]
Chinese Medicinal Plants from the Pen Ts'ao Kang Mu. Peking, 1936.

Sivin, Nathan
Chinese Alchemy: Preliminary Studies. Cambridge, Mass., 1968.

T'ang Shen-wei 唐愼微 et al.
Ch'ung hsiu Cheng-ho ching shih cheng lei pei yung pen-ts'ao 重修政和經史證類備用本草 (1249). Reprint from the first edition, Peking, 1957.

Ware, James R.
Alchemy, Medicine, Religion in the China of A. D. 320: The Nei P'ien of Ko Hung (Pao-p'u tzu). Cambridge, Mass., 1966.

Wu Ch'i-chün 吳其濬
Chih-wu ming shih t'u k'ao 植物名實圖考 (this and the more detailed preliminary draft below were first printed in 1848, two years after the decease of its author). Shanghai, 1919.

——————

Chih-wu ming shih t'u k'ao ch'ang pien 長編 . Shanghai, 1919.

8

Man as a Medicine: Pharmacological and Ritual Aspects of Traditional Therapy Using Drugs Derived from the Human Body

William C. Cooper and Nathan Sivin

Traditional Chinese medicine strove to treat the whole person rather than his isolated parts, and to think of him in relation to his emotional sphere and physical environment. This ancient diagnostic and therapeutic approach in many respects anticipates that of the sophisticated modern doctor. But the criteria that the Chinese physician used to verify effectiveness of a treatment were drastically different from those of modern scientific medicine—although perhaps less easily distinguishable from the attitudes and rules of thumb still current in the broad areas of medical application in which doctors are still willing to credit anything that seems to work regardless of the demands of systematic verification. More to the point in attempting to understand Chinese practice are its criteria of disproof, which appear to have been particularly permissive. The relation between concept, observation, and method did not ordinarily necessitate the rejection of therapeutic procedures whose results could as well have been obtained by no treatment at all, nor the simplification of prescriptions the efficacy of which would not have been affected by the elimination of most of their ingredients. Perhaps the most fundamental limiting factor was the lack of concern to distinguish those therapeutic measures that are specifically indicated for particular diseases. As we shall show, drug therapy is a hodgepodge of specific remedies, treatments of specious specificity, panaceas, partly or wholly deritualized rituals, and thaumaturgical tours de force.

Established remedies were indeed rejected, and worthy new ones accepted, as one pharmacopoeia after another was compiled through history. That even conceptual and therapeutic revolutions occurred from time to time should not be denied a priori simply because few historians have thought of looking for them. Still, Chinese medicine shared with other premodern medical systems the lack of steady methodological pressure for changes of these kinds. Its dynamic view of the body and its disorders, far from being empirical, was built upon metaphysical concepts (yin-yang, *wu-hsing* 五行 or Five Phases, and others) that experience could not discredit.

Since Chinese anatomic and physiologic concepts were not consistently and rigorously tested against a reality that resembles our own, we are naturally tempted to ask whether the system's ability to survive really had little or nothing to do with rigorous experiential testing, whether it was merely the result of luck or unsystematic common sense that the traditional physician was able to cure his patients with some regularity. To answer "yes" is to see the growth of Chinese medicine as in the last analysis random, and thus to give up the possibility of understanding its gradual historical definition against a background of innumerable alternative approaches. The literature of Chinese medicine is, after all, a partial record of which possibilities were chosen while others were rejected, and why. The serious and attentive study of these rationales for historical choices has hardly begun, but there is no longer room for doubt that they formed conscious, consistent, coherent patterns.

With that much said, it would still be unwise in exploring the character of any medical system (including our own) to ignore the fact that between half and three-quarters of man's cases of illness are not affected in a specific way by drug treatment, either because they would run their courses in any case or because they are psychosomatic. It is not true, as many laymen think, that psychosomatic disorders do not manifest themselves physically, or that drugs do not affect them under any circumstances.[1] The Chinese physician, with his prestige

[1] See T. R. Harrison, R. D. Adams, I. L. Bennett, Jr., W. H. Resnik, G. W. Thorn, and M. M. Wintrobe, eds., *Principles of Internal Medicine* (3rd ed., New York, 1958), pp. 230–231.

and his impressive therapeutic armamentarium, was an excellent practical psychotherapist. He used a mixed therapy consonant to his broad approach to the patient. Calisthenics, massage, drugs, acupuncture, moxibustion, incantations, and magical amulets were among the measures he could combine as he chose. The medical efficacy of the first three (at least as genres) is familiar to modern medicine. It is inevitable that their combined use should have led to giving the last four credit for physiological potency which it has yet to be rigorously proved they deserve. But while this may be a partial explanation, it does not account for the ability of all seven to compete successfully for survival, to satisfy the clinical and social demands made upon them, for at least two millennia. We must therefore take seriously that truism of psychosomatic medicine which says that the belief tends to lead by obscure paths to its own validation.

The last statement is not meant to deny the likelihood that continued scientific experimentation will reveal a clear physiological basis for at least some of the curative powers of needle and moxa therapy. Modern medicine remains too ignorant of the part of the nervous system that lies near the surface of the skin to be sure that action upon it is not partially responsible for the reported efficacy of acupuncture and moxibustion. There have been many experiments in the past, and they have yielded inconclusive, contradictory, or irreproducible results, but this is perhaps inevitable for investigations meant not to comprehend but to vindicate and justify. Our concern is not to predict the outcome of investigations yet to be made, but merely to apply the same skeptical standard any physician would demand for a new therapy. At the same time, when studying Chinese medicine one spends so much time in the vicinity of the border between the physiological and the psychological that it is impossible not to recognize that the two regions are sensibly superimposed. To assume that the ability of a placebo to cure sciatica in a double-blind experiment is due to *either* physiological *or* psychological mechanisms is surely to miss the point.

Our purpose is to begin the serious examination of the processes by which certain methods and agents without demonstrable pharmaco-dynamic efficacy under the conditions of modern medical experimentation were retained in use by Chinese physicians. In order to focus as squarely as possible on this simple issue, we will avoid the complications inherent in comparing one discrete therapy with another and deal with the problem of medical effectiveness as it arises in considering a group of drugs alone. The mixed-therapy problem recurs nonetheless, because as we shall demonstrate, there was no such thing in Chinese drug therapy as an empirical tradition based on pharmacological properties alone. Magic and ritual play too large a role in drug formulas to ignore, and clinical experience was always shaped and interpreted in terms of theory. But circumscribing our subject matter does at least allow us to confront the problem in a more manageable form.

The heterogeneous character of Chinese materia medica begins to make sense only when we realize that the tradition of the great pharmacopoeias (which is parallel to that of the great theoretical and therapeutic treatises) is anything but folk medicine. It is a comprehensive and highly rational body of theory and practice held together by the most fundamental concepts of classical natural philosophy, which have been refined, modified, and elaborated to incorporate the phenomena of health and sickness. In form, style, and level of abstraction it has belonged to the "great tradition" of China's tiny educated elite. But it depended for much of its vitality upon a symbiotic relation with the folk medicine of the overwhelming majority of peasants, merchants, artisans, and others (including the uneducated wives of the elite) whose world was not only much more intellectually restricted but full of personal forces, spirits and ghosts, which brought and took away sickness and other visitations of fate (the "small tradition"). Despite endless change in the social patterns of medicine and in the points of division between the great and small traditions over the centuries—and indeed the constant presence of physicians at home in both—the contrast between the spiritualistic world view of folk medicine and the abstract speculative cosmology of classical medicine is always discernible.

The long series of books on materia medica that survive today were written to be classics. The progenitor of the line was attributed to a legendary demiurge-emperor and given canonical status. These books were meant to be read by the tiny minority of literate physicians who could be expected to transmit and carry on the tradition. But we know well enough that before a drug could be rationalized and canonized, so to speak, it had to prove itself through long use by village practitioners, healers who journeyed into the mountains to gather their own simples, illiterate and semiliterate doctors unknown outside their ward of the city, operators of therapeutic cults, and wise men and women whose training included more ritual and magic than pharmacognosy.[2] It is hardly to be expected that in this preliminary screening and testing the operative criteria should have had much in common with those employed by a modern Swiss pharmaceutical laboratory. A decisive difference is that Chinese doctors did without the idea of the experimental control, despite much versatility in working out approaches to drug experiment. We find reports of trials upon animals, but the rule, as in other traditional systems, was to observe human patients.[3] What, then, was the interplay of empirical, psychosomatic, ritual, and theoretical factors in the Chinese doctor's evaluation of his clinical experience?

[2] We do not use the words "ritual" and "magic" quite interchangeably. "Ritual" is a wider category, which includes procedures in which spiritual agencies are persuaded to bring about the desired change as well as those in which the operator brings them about directly by application of magical principles or laws. We refer occasionally to the mode of thought that underlies both categories as "symbolic." Although in the West only the first kind of procedure would be considered religious, Taoism made use of both. Cf. John Middleton, ed., *Magic, Witchcraft, and Curing* (American Museum Sourcebooks in Anthropology, Q7; Garden City, New York, 1967), p. ix.
[3] On animal experimentation, see Ho Ping-yü and Joseph Needham, "Elixir Poisoning in Mediaeval China," *Janus*, 1959, *48:* 227–228, reprinted in Needham, *Clerks and Craftsmen in China and the West. Lectures and Addresses on the History of Science and Technology* (Cambridge, England, 1970), p. 322. On behavioral experiments, see the article of Miyasita Saburō in this volume.

This is a central question and can best be answered by examining, in the cases of a number of drugs, what he believed their curative powers to be. We do not attempt to provide a definite answer to this basic question in our very preliminary exploration. Our aim is merely to estimate the effectiveness of a group of traditional remedies according to the *relatively* culture-free criteria of modern medical science, to explore the problems that arise in making such estimates, and to suggest in a very tentative and speculative way the implications of our findings for the next stages of investigation.

There is a very real confusion in the notion of clinical effectiveness that must not be overlooked. The vocabulary of modern medicine is very poorly adapted to dealing with cultural predisposition to disease and its corollary, cultural predisposition to the cure of disease. We have experienced much frustration while trying to express intelligibly the idea that ancient Chinese physicians cannot be considered negligent for having failed to distinguish a cure effected by a highly specific drug (in our eyes) from relief due, say, to writing a magical diagram, burning the paper to ashes, and having the patient wash the ashes down with water drawn from a well at dawn. When a given physician shows a preference for the specific drug we tend too lightly to credit him with an inexplicable scientific instinct that puts him ahead of his time, rather than recognizing that his criterion may have been a matter of social identification or even snobbery; magical diagrams and incantations were seen by many as vulgar, fit only for commoners. Any general measure of effectiveness has to include both the specific and the unspecific cure, just as in the ordinary practice of the modern physician his necessarily rough-and-ready personal estimate of the worth of a new drug will inevitably depend partly on cures of his patients that could as well have been brought about by a sugar pill or by time alone. This does not, granted, present the doctor of today with a problem of great urgency. But any study such as ours would be valueless if it made the error of imagining that one can take as a measure of the total clinical effectiveness of Chinese drugs pharmacologic properties that are defined, known, and used within the very different therapeutic ambiance of modern medicine. We can use those properties as a tool with which to separate specific remedies from those that are effective to some unknown but far from negligible degree in the special circumstances of Chinese culture. But we see this separation primarily as prerequisite to the serious examination of those therapeutic procedures that are *unsound* by the criteria of current pharmacology, and their undeniable ability to survive alongside those that modern science considers sound.

The Chinese pharmacopoeias (*pen-ts'ao*) contain literally thousands of descriptions of simple medicines of vegetable, animal, and mineral origins. Chinese drugstores are replete with the widest imaginable assortment of remedies from ginseng root to dried salamanders. It would be no trivial task to choose a small representative sample of these medicinals and attempt to relate scientifically their active constituents to the disorders for which they are prescribed. In most cases we know too little about what those active constituents are. Therefore we have chosen to consider man as a medicine, and to examine what the Chinese called human drugs. The chemical composition of man and of his various excreta, appendages, and secretions is mostly known, and their therapeutic effectiveness, or lack of it, can be objectively, if approximately, estimated without reference to their Chinese context, on the basis of modern practice in a variety of cultures. Once we have isolated some of the areas in which known scientific efficacy and traditional therapeutic claims do not correspond, we shall have taken a large and indispensable step toward the eventual rigorous investigation of cultural and psychosomatic factors.

It has probably occurred to the reader that human drugs are not representative of the entire range of Chinese pharmacognosy. This is quite true; there are many remedies for most diseases, and the human drug prescriptions, for obvious reasons, have not often been the most popular. But they were certainly much used over the centuries. Characteristics that they reveal in this chapter—the inundation of clearly useful remedies by many others that appear useless, the ready recourse to magic and ritual, and the frequent complexity of preparation far disproportionate to predictable therapeutic benefit—are common throughout the pharmacopoeias.[4]

Most prior investigations have aimed at the exploitation of Chinese drugs in modern medicine, or occasionally at a vindication of the value of traditional medicine. They have naturally either ignored claims for specific effectiveness that modern medicine does not support, or cited them in order to denigrate the ancient system as a whole. In order to comprehend the nature of the classical system of medicine, which is our concern, it is imperative to reexamine the insupportable claims (which are usually ignored) with the same attention as the sound claims. It is well known that drugs of the ephedra family 麻黃, mainly because of their alkaloid ephedrine, would have proved their value in producing vasoconstriction, bronchodilatation, contraction of the uterus, and as central nervous system stimulants; but they were also prescribed for minor epidemic febrile diseases 天行熱病 and for black pustules in smallpox 痘瘡倒靨.[5] The concentrated sex hormone preparations, recently described by Lu Gwei-djen and Joseph Needham in what must be considered the first serious historical study of Chinese human drugs, may have had real value in some cases of impotence and other sexual disorders; but we also find them recommended for general inability to swallow food 噎食反胃 and

[4] For a wide selection of prescriptions, see A. Chamfrault and Ung Kan Sam, *Traité de médecine chinoise* (5 vols., Angoulême, 1954–1963), Vol. IV (*Formules magistrales*). The materia medica in Vol. III, which are said to be represented in current practice in the Chinese community of Vietnam and to have been tested by Chamfrault's collaborator, include a number of human substances: fingernails, hair, teeth, excrement, urine, and placenta (III, 19–22). Ritual elements over the whole range of materia medica will be discussed in a study of Taoism as a mediator between folk medicine and classical medicine, to be presented by one of us (N.S.) at the Second International Conference on Taoism (1972).

[5] In the Great Pharmacopoeia of 1596 (*Pen-ts'ao kang-mu* 本草綱目; reprint of the 1930 Wan yu wen-k'u 萬有文庫 ed., 6 vols., Peking, 1954), *15*: 66–67.

fevers following ingestion of alchemical elixirs, which usually contained heavy metal compounds 服丹發熱.[6]

The history of Chinese drug therapy is essentially the sum of the conceptual and social processes by which every medicinal substance was assigned every one of its applications. To pick and choose according to which curative claims strike a chord in the mind of the modern physician can yield much practical and interesting information—much of it a badly needed corrective to the intellectual provincialism of many Western doctors—but it will not yield a history of Chinese medicine.

For each of the eight substances derived from man that is considered in this chapter, we present information on its preparation and use from the Great Pharmacopoeia of 1596 (*Pen-ts'ao kang mu* 本草綱目 of Li Shih-chen 李時珍), still one of the traditional physician's standard reference books. Our commentary at the end of each section discusses the known chemical composition of the human substance (as well as that of other constituents of the standard prescriptions) and takes up the question of whether the Chinese applications can be related positively to any medicinal benefit known to modern science.[7]

[6] Ibid., *52:* 96; Lu Gwei-djen and Joseph Needham, "Medieval Preparations of Urinary Steroid Hormones," *Medical History* (1964), *8:* 101–121.

[7] The section on human drugs in the Great Pharmacopoeia (*52:* 80–106) includes the following substances in addition to those treated in this paper: hair collected from a comb; dandruff; earwax; dirt from the kneecap*; excrement; placental excrement (? 小兒胎尿)*; urine; urine sediment; purified and concentrated urine preparations; urinary calculi*; gallstones*; soft deposit scraped from teeth; perspiration*; tears*; breath*; the yin soul of a hanged person (a material object resembling pine charcoal dug out of the ground beneath the body shortly after death)*; facial hair*; pubic hair; placenta; liquid from a placenta that has decomposed after long burial*; umbilical cord*; penis*; gall; flesh*; and a foreign nostrum prepared from the body of a holy man who has lived on honey during his last days and whose corpse has been immersed in honey for a hundred years* (so at least goes the Chinese description of that great Western nostrum mummy 木乃伊). Our selection is to some extent arbitrary, dictated by a desire to keep this paper within manageable proportions, but we have

In the Great Pharmacopoeia's systematic article on each medicinal substance, there are two rubrics that have to do with curative virtue.[8] The first, "Physiological Activity and Diseases Cured" (*chu chih* 主治), is a list mostly excerpted from earlier classics of pharmacognosy. The function of this section is more historic and theoretic than practical. It demonstrates the breadth of application of the drug but gives inadequate information about compounding and administration, merely specifying the form in which it is prepared for various uses. To depend upon it would mean remaining ignorant of other major ingredients, one of which may be, in the eyes of modern physician, the active ingredient. The section "Appended Prescriptions" (*fu fang* 附方), on the other hand, gives a series of detailed preparations, with indications of dosage, for specified disorders. Although many of these also come from early sources, they were included in Li's compilation simply because he believed them useful in the practice of his own time.

---

included a majority of the substances that we have reason to believe were widely used. No prescriptions are given for many of the human drugs (those followed by an asterisk in the list above), and Li Shih-chen even condemns the use of a few of them (especially human gall and flesh). This is not surprising in view of the well-known Confucian inhibition against mutilating one's body, but it is clear that some doctors did not consider it an absolute limitation. We have omitted urine and its derivatives to avoid going over the same ground as Lu and Needham in the paper cited earlier.

The medical discussions in the following sections are primarily the responsiblity of W. C. Cooper, and translation of prescriptions and interpretation of ritual elements that of N. Sivin. The remainder of the paper represents the consensus of both collaborators. The authors are grateful to Arthur M. Kleinman, M.D., for encouragement and counsel that led to a revision in the approach of this investigation.

[8] For the divisions of *Pen-ts'ao kang mu* see E. Bretschneider, "Botanicon sinicum. Notes on Chinese Botany from Native and Western Sources," *Journal of the North China Branch, Royal Asiatic Society*, 1881, n. s., *16:* 56–63, and Lu Gwei-djen, "China's Greatest Naturalist; A Brief Biography of Li Shih-chen," *Physis*, 1966, *8:* 383–392; Chang Hui-chien, *Li Shih-chen—Great Pharmacologist of Ancient China* (Peking, 1960). For the pharmacological tradition as a whole, see P. Huard and M. Wong, "Évolution de la matière médicale chinoise," *Janus*, 1958, *47:* 3–67, which should be corrected in detail by recourse to reference works in Chinese, such as Lung Po-chien 龍伯堅, *Hsien-ts'un pen-ts'ao shu lu* 現存本草書錄 (A bibliography of extant pharmacopoeias; Peking, 1957).

The authority of his book has given them currency up to the present day.[9] It is therefore the generally more limited range of claims in the "Appended Prescriptions" which is reproduced here, with each prescription fully translated.

Perhaps the most difficult methodological issue faced in the course of this investigation has to do with the identification of Chinese disease entities.[10] The Chinese system of diseases was defined out of experience with the same infections, deficiencies, toxifications, and so on, as our own. The demarcation of one disorder from another, and their mutual relations, are very different, since they grew out of a different system of concepts and ways of looking at the body. The distinction between symptom and disease also tends to be vague. Consequently, most of the Chinese names of diseases can be translated only loosely into an English equivalent, and many not at all.[11]

[9] It has remained one of the most-used reference works of every traditional doctor, whether in China or in Chinese communities overseas—although the internal situation may have changed since the Cultural Revolution. In order to demonstrate its currency I will merely note the publication figures in my copies of the two major postwar editions. By the end of the sixth printing (1959) of the Commercial Press edition of 1954, 32,500 copies had been printed. The fourth printing (1963) of the People's Hygiene Press edition of 1957 notes 36,500 copies to date (N.S.).

There is of course the further question of whether human drugs are still being employed in China, or whether the part of the traditional pharmacopoeias that describes them is simply ignored. Whether individual doctors are using them can be determined only by field research, but there is no doubt that as of 1961 their use was still being taught. *Yao-ts'ai-hsueh* 藥材學 (Pharmacognosy; Peking, 1961, reprint, Hong Kong [,1970?]), compiled by the Nanking College of Pharmacy, is an important recent handbook of the drug resources of traditional Chinese medicine as currently practiced. Its 634 carefully chosen medicinal substances include 82 of animal origin. Among these are five human drugs: placenta 紫河車, fingernails 人退, ashed hair 血余, urine 童便, and urine sediment 人中白 (pp. 1270–1273).

[10] For our approach to the identification of medicinal substances and disorders, see N. Sivin, *Chinese Alchemy: Preliminary Studies* (Harvard Monographs in the History of Science, I; Cambridge, Mass., 1968), pp. 272–321. Here we use a less chronologically restricted range of sources, since our concern is with the post-Ming understanding of the terms we study.

[11] We are aware of the inconsistency involved in translating what we admit is untranslatable. We can see only two alternatives. The first is not to translate at all, which would deprive the reader of the evidence upon which

Instead of a term for tuberculosis, we have several Chinese diseases, each of which includes tuberculosis as well as an assortment of other disorders whose symptoms fall into one or another category. The Chinese entities may be completely unrelated according to traditional nosology, even though we would say that many or most occurrences of any of them would be diagnosed by a modern physician as pulmonary tuberculosis. This defective correspondence places limits on the absolute validity of any study of this kind. We have attempted to minimize its effect by taking it explicitly into account. The name we give for each disorder usually represents a straightforward rendering of the Chinese term. Only when the latter is extremely specific or clearly technical is the corresponding modern entity given as its translation. "*Huang tan* 黃疸" is neither etiologically nor symptomologically identical with jaundice, but the first graph means "yellow" and the second is the name of jaundice and related disorders which the Chinese did not conventionally distinguish from it. So long as the term is to be translated at all, "jaundice" is the only feasible simple rendering. We have taken special pains to avoid mechanically piling up the dictionary definitions of one character after another. "Unexpelled placenta" may lack the exotic charm (and the philological authenticity) of "foetus garb undescended," but it makes the pathological point unambiguous, and

---

the translations are based, and greatly lessen readability. The second is to coin new English terms, possibly from Greek and Latin roots. That is a philologically superior solution, and new systems of nomenclature are now being worked out independently by Needham and Lu in England and Manfred Porkert in Germany. We have not followed their lead partly because of the restricted scope of this paper and partly because there will be more than enough confusion caused by the competition between these two new systems without our contributing a third. But the reader should remain aware that such terms as "fire" and "wind" are functionally defined and do not refer merely to the phenomena of everyday experience.

"*Wu-hsing*," which we translate by "Five Phases," is rendered by Porkert as "Five Evolutive Phases" and by Needham as "Five Elements" (although he points out that this is potentially misleading). *Wu-hsing* is a fivefold division which can be applied to any cyclic process or configuration. In medicine it is preponderantly temporal rather than spatial and energetic rather than material.

renders a technical term by a corresponding technical term. If the name we give for a disease does not correspond reasonably to an entity in modern medicine, we suggest possible equivalents, or at least outline the basic symptoms so that the reader may assess for himself or herself the problems of identification.[12] "*Yin-yang i ping* 陰陽易病," for instance, is a general name for the pathological effects of what originally must have been a taboo violation (see pp. 220–221). We can do no better in this case than translate the sense of the words as they would be understood in context by a Chinese physician: "yin-yang exchange disorders." The additional data we adduce to fill out our explanations come from sources easily accessible to the traditional practitioner, so that our remarks may reflect his understanding.

In each item translated below, the first paragraph gives our identification of the disorder, and in most cases supporting data and concise references to the literature. The second paragraph is our verbatim translation of the prescription from *Pen-ts'ao kang mu*. Where individual medical commentary on the therapy is needed, it appears as a third paragraph.

Key to Abbreviations

C

Chu ping yuan hou lun 諸病源侯論
On the origins and symptoms of diseases, 610
Ch'ao Yuan-fang 巢元方
Peking, 1955

H

Ch'ung hsiu Cheng-ho ching shih cheng lei pei yung pen-ts'ao 重修政和
經史證類備用本草
Revised convenient pharmacopoeia, with classifications verified from the classics and histories, 1249

---

[12] Sometimes the designation of a disorder in the Great Pharmacopoeia is idiosyncratic, even differing from the term given in Li's source. This is partly due to a stylistic peculiarity. All the names of diseases for which prescriptions are given under a particular drug are expressed in the same number of characters, either two or four, with the latter predominating.

T'ang Shen-wei 唐愼微
12 vols., Peking, 1957

K

Chung-i-hsueh kai-lun 中醫學概論
Summary of Chinese medicine, 1959
Nanking College of Chinese Medicine (ed.)
Reprint, Hong Kong, 1969

L

Pen-ts'ao kang mu 本草綱目
Systematic pharmacopoeia, 1596 ("Great Pharmacopoeia")
Li Shih-chen 李時珍
6 vols., Peking, 1930, reprinted 1959

S

Ch'ien chin fang 千金方 (*Pei chi ch'ien chin yao fang* 備急千金要方)
Prescriptions worth a thousand, 650/659
Sun Ssu-mo 孫思邈
Tokyo, 1849 (Edo Igaku 江戶醫學), reprint, Taipei, 1965

T

T'ai-p'ing sheng hui fang 太平聖惠方
Imperial Grace formulary of the Grand Peace era, 992
Wang Huai-yin 王懷隱
2 vols., Peking, 1959

W

Ping-yuan tz'u tien 病源辭典
Dictionary of semeiology
Wu K'o-ch'ien 吳克潛
Kowloon, 1965

Y

Ku-tai chi-ping ming hou su i 古代疾病名候疏義
Glosses on ancient names and symptoms of diseases
Yü Yun-hsiu 余雲岫
Peking, 1953

I. Human Hair 髮髲

Li Shih-chen notes that hair is taken from the centers of the crowns of the heads of healthy males twenty years old or so with sound complexions. The hair is washed with soap-bean water, sun-dried, and ashed in a sealed pot.

1. (Identification of disease:) Stones of the urinary bladder, causing painful and rough urination 石淋痛澀. One of a series of disorders (*lin* 淋) whose symptoms are abnormal urination, caused by a pneumatic deficiency in the kidneys and hot factors in the bladder (*C 14:* 83).

(Translation of prescription:) The hair is roasted to ash without physical decomposition 存性 and ground to powder. One *ch'ien* (4 g) is taken with well water.[13]

2. Yellow disease due to cold factors 傷寒黃病 (= infectious hepatitis). This belongs to the great class of infectious febrile diseases due to cold factors and is distinguished by yellow coloring of the face, aching throughout the body, and fever (*C 12:* 70).

Hair is roasted and ground. A standard spatula-full is taken three times a day with water.

3. Unexpelled placenta 胎衣不下

Hair from the head and loose hair taken from a comb are straightened out, knotted, and held in the mouth (?撩結口中).

4. Possession of a small child by the *ch'i* (pneuma) of an offended spirit 小兒客忤 (= attacks of acute anxiety). Caused by the child's seeing a stranger or something unexpected while his own vital spirits and *ch'i* are debilitated, and manifested in vomiting, upper abdominal pain, and seizures which resemble those of an epileptic except for the absence of eye movement (*C 46:* 248). Marjorie Topley has shown in her study of disease conceptions in modern Hong Kong that in folk belief this disease state involves the flight of the child's yang soul (*hun* 魂) from its body. The use of hair as a binding agent in magic is well known.[14]

[13] The equivalents given throughout for weights and measures are those official in the late sixteenth century.
[14] Marjorie Topley, "Chinese Traditional Ideas and the Treatment of

Brought about by seeing a stranger: Take ten hairs from the crown of the head of the person who has come, cut off a bit of the child's belt, mix them, roast, and grind to powder. Mix the powder with milk and give it to the child to drink, and it will get better.

5. Quick stomach with aching 急肚疼病 (= colic?)

Use thirty hairs from the head of the patient, and after roasting have him take them with wine. Then mix powdered mustard seed with water and pack it into his navel. When his sweat comes down like rain the disorder will subside.

6. Felons and suppurating abscesses 瘭癌惡瘡

Take two *ch'ien* (8 g) of the ash of living hair with rice-cooking water. Alternately take three *fen* (28 g) of the ash of living hair mixed dry with two *fen* of ashed soap-bean tree thorns 皂莢刺 (*Gleditschia sinensis*, Lamarck) and one *fen* of the powdered tuber of *Bletilla striata*, (Thunberg) Reichenbach 白芨. Apply the powder dry 乾摻. This compound is sometimes mixed with pig bile.

Remarks: The use of soap-bean tree thorn ash with hair might prove to be of minor therapeutic benefit. *Gleditschia sinensis* contains high concentrations of saponins—glycosides that are highly surface-active, and foam when mixed with water. One might postulate a detergent and cleansing effect upon the abscess if moisture from the skin wet the powder, although the dry powder itself (which the prescription designates) would not be effective as a cleansing agent.

COMMENTARY

The six entities treated with human hair are entirely unrelated, representing as wide an assortment of afflictions as in any of the larger categories which follow. No linkage can be detected among these diseases that might relate them to each other either theoretically, empirically, or magically.

Disease: Two Examples from Hong Kong," *Man*, 1970, *5:* 429–434. On hair as a binding agent, see Edward H. Schafer, *The Golden Peaches of Samerkand. A Study of T'ang Exotics* (Berkeley and Los Angeles, 1963), pp. 193–194. For various technical meanings of "*ch'i*" in Chinese science, see Sivin, *Chinese Alchemy*, index, s. v.

Pains were taken to obtain hair from healthy scalps and to clean the hair thoroughly before roasting or ashing it. The use of hair from young healthy males has strong yang implications; the strength and purity of the active agents would be assumed. The "notion of binding, tying up, and holding fast," which governs much magical application of human hair, is no longer apparent in the highly rationalized *Pen-ts'ao kang mu*, but is implicit, for instance, in the fourth prescription.

In human hair the concentrations of calcium, phosphorus, iron, and copper are high, and trace metals known to be essential in human nutrition are present, such as cobalt (the essential ligand of cyanocobalamin, or vitamin B_{12}, used in treatment of pernicious anemia); manganese, a possible cofactor for the mitochondrial respiratory enzyme arginase; and zinc, a necessary cofactor for the enzymes carbonic anhydrase and carboxypeptidase.

Hair has a relatively high content of the sulfur-containing amino acid cystine. Arsenic is known to be concentrated in hair and nails during chronic arsenic ingestion in man; hair and nails are often used in tests to determine whether a person has been chronically poisoned with arsenic. Hair may also concentrate other trace metals such as aluminum, lead, and uranium.

There are no significant color-related differences among the chemical components of hair except in the case of lead, which is present in black hair in much higher concentration than in brown hair. Chinese hair is invariably black. The concentration of lead in female black hair is 28.4 mg/100g; one would not expect male black hair to contain less.[15] There is a possibility that chronic use of black human hair ash by a given patient (in immensely larger amounts than Li Shih-chen seems to contemplate) might lead to chronic lead poisoning, with its accompanying gastrointestinal symptoms, anemia, and dental and bone changes.

On the basis of the chemical composition of human hair,

[15] P. L. Altman and D. S. Dittner, eds., *Biology Data Book* (Washington, D.C., 1964), s. v.

there is no reason, pending a better understanding of the biochemistry of trace elements, to believe that any therapeutic benefit would result from its use in any of the ailments described.

II. Nails 爪甲

1. The Great Pharmacopoeia's first procedure is not from medical literature proper but from Taoist ritual for expelling the Three Corpse-Worms 三尸, the chief of the "inner gods" who are to the individual microcosm what the celestial bureaucracy is to the cosmos.[16]

The devotee cuts his fingernails on the seventeenth day of each sixty-day cycle and his toenails on the thirty-first. Then on the sixteenth day of the seventh month he roasts the trimmings to ash and takes them with water. One source says that on the fifty-first day of each cycle the Three Worms visit the two hands, so then the fingernails are cut; on the thirty-first day the Three Worms visit the two feet, so then the toenails are cut.

2. Foot-*ch'i* 消除腳氣 (= beri-beri)[17]

On the third day of the cycle of twelve days, cut the fingernails and toenails, slightly impinging upon the flesh.

3. Disorders due to wind agents entering an open wound (see p.229). Two prescriptions are given, of which we translate one.

The trimmings of all the fingernails and toenails are sautéed in sesame oil, ground, and taken in hot wine. Sweating indicates that remission is imminent.

4. Yin-yang exchange disorders 陰陽易病. A general term for illnesses contracted through sexual intercourse with someone who has just recovered from them. They are diagnosed by the

[16] R. Hoeppli, *Parasites and Parasitic Infections in Early Medicine and Science* (Singapore, 1959), pp. 62–63. The Three Worms are illustrated in the anonymous and undated Grand Superior Canon for Expelling the Three Corpse-Worms and Nine Worms and Maintaining Vitality 太上除三尸九蟲保生經 in vol. 580 of the Taoist Patrology 道藏. The role of fingernail parings in Taoist ritual will be examined in detail elsewhere (see note 4).

[17] Lu Gwei-djen and Joseph Needham, "A Contribution to the History of Chinese Dietetics," *Isis*, (1951 submitted 1939), *42*: 13–20.

appearance of a yang pulse on the wrist under conditions in which a yin pulse is normally read, and vice versa (*W* 627; cf. R.H. van Gulik, *Sexual Life in Ancient China,* Leiden, 1961, p. 152; Akira Ishihara and Howard S. Levy, *The Tao of Sex. An Annotated Translation of the Twenty-eighth Section of "The Essence of Medical Prescriptions (Ishimpō),"* Yokohama, 1968, pp. 137–138.) Marjorie Topley has found in Hong Kong folk medicine a very similar disease of males caused by violating the prohibition against intercourse within 100 days after childbirth. Some such taboo origin for Li Shih-chen's much more general disease class is quite conceivable; 100 days is a very common ritual interval in China.[18]

The nail trimmings of all twenty fingers and toes are ashed with a piece out of the seat of a pair of trousers (a male's if the patient is female and vice versa) and taken in three portions with warm wine.

5. Swollen belly in infants 小兒腹脹. Probably ascites, but the Chinese thought of it as an accumulation of wind or cold *ch'i* (*C 47:* 250).

The fingernail trimmings of the child's father and mother are ashed and smeared on the mother's nipple, and the child is suckled.

6. Urine retention due to retroverted bladder 小便轉胞. Difficulty in urination due to prolonged voluntary urine retention (while traveling or after feasting), in some cases acute or even fatal (*C 14:* 85).

The patient's nail trimmings are ashed and taken with water.

7. Abnormal urination in men and women 男女淋疾. A general term for disorders in which the urine is discolored, contains deposits, or is voided with difficulty. They are caused by a *ch'i* deficiency in the kidneys and hot factors in the bladder (*C 14:* 83).

Treated as above.

8. Hematuria 小便尿血

Half a *ch'ien* (2 g) of the patient's fingernail trimmings are

[18] Topley, "Chinese Traditional Ideas," references to *so-lò* on pp. 424 and 429.

ashed with 1½ *ch'ien* of his hair and the residue ground to powder. One *ch'ien* of the powder is taken at a time with warm wine on an empty stomach.

9. Hematuria in pregnant women 妊婦尿血

Fingernail trimmings of the patient's husband are ashed and taken with wine.

10. Unexpelled placenta 胞衣不下

Trimmings of the woman's fingernails and toenails are ashed and taken with wine. Then, while a strong woman holds her up, a bamboo tube is manipulated downward in front of her chest, making the motion of expelling the placenta (? 即令有力婦人抱起. 將竹筒于胸前趕下).

11. Painful ulcerated hemorrhoids 諸痔腫痛

Pack a silkworm cocoon full of fingernail trimmings from a man, and bind it with hairs from the crown of a boy's head. Roast it, avoiding physical decomposition, grind it to powder, mix it with honey, and spread it on the ulceration. In addition, pagoda tree fruit 槐子 (*Sophora japonica,* L.) which has been treated in ox bile, taken daily, is marvelously effective.

12. Needles stuck in the flesh 鍼刺入肉

For needles broken off in the flesh, and wood or bamboo splinters. Scrape off some fingernail to form a powder and pound it with wild jujube flesh 酸棗 (*Ziziphus vulgaris var. spinosum,* Bunge) to form a disintegrated mass. Spread it on the lesion, and the next day the splinter will come out without fail.

13. Flying gossamer in the eye 飛絲入目

Scrape off some fingernail to form a powder, moisten it with saliva and, using the fingertip, touch it to the eye. The gossamer will collect and can be lifted out.

14. Foreign bodies in the eye 物入目中

With a knife scrape off and powder some of the left-hand fingernails. Dip rush pith 燈草 (*Juncus effusus,* L., *var. decipiens,* Buchen.) in the powder and apply it to the canthus.[19] After three times the foreign body will come out.

---

[19] The character "*i* 瞖," which means a cataract or other physical obstruction to sight, is obviously out of place here. In its stead we read the visually similar "*tzu* 眥" (canthus).

15. Cataract due to smallpox 瘢痘生翳 (= acute eye infection due to smallpox)

Treated as above.

16. Eye diseases in general. Two further prescriptions for particular eye diseases are omitted.

Pick up some powdered fingernail by rubbing horsetail 木賊 (*Equisetum hyemale*, L.) across the nail, and grind with equal parts of cinnabar until uniform.[20] Using dew for moisture, make pills the size of a mustard seed and put one into the eye for each treatment.

17. Chronic bloody diarrhea 積年瀉血

Incurable by other drugs. Take $2\frac{1}{2}$ *liang* (93 g) each of fingernail trimmings sauteed until black, and musk 麝香; 3 *liang* of roasted dried ginger 乾薑; and 1 *liang* (37 g) each of crystallized alum 白礬 that has been desiccated, and worn-out thin leather 敗皮巾 that has been roasted to ash. Powder. For each dose take 1 *ch'ien* (4 g) with rice-gruel, twice a day.

18. Nosebleed 鼻出衄血

With a knife scrape off some fingernail to form a fine powder. When it is blown into the nose the epistaxis will stop. Verified by trial.

COMMENTARY

Of the eighteen entities under the heading "Nails," one (beri-beri, item 2) is treated by manipulation, within a time cycle, of the patient's own nails; nails are not used as a drug in this instance. Beri-beri is a nutritional disease caused by dietary deficiency of thiamine (vitamin B_1). In China, this disease usually resulted from the practice of eating polished rice almost exclusively, inadequately supplemented with thiamine-containing foods. Absolutely no benefit would result from cutting the nails; impingement upon the flesh would only serve to aggravate the paresthesias in the feet (burning, etc.) caused by the peripheral neuropathy of beri-beri. Sympathetic magic may explain the origin of this treatment, but the associations

[20] The cuticle of the horsetail is highly siliceous, and has been much used as an abrasive.

are not explicit in our source. Since nails, in this instance, were not used as a drug, we can only look for a ritual explanation.

Item 1 involves Taoist ritual, not medicine, and we shall not discuss it further. Item 4 is a class of diseases which have nothing in common but mode of transmission. This pathological entity, as we have noted, may have been abstracted or even generalized from a popular taboo. Item 16 (eye diseases in general) must be omitted from discussion because it is too broad a category.

Fourteen disorders now remain: three related to the eyes, four dealing with the urinary tract, two varieties of external open wounds (3 and 11), two types of bleeding disorders (17 and 18), a gastrointestinal disorder in infants (5), unexpelled placenta (10), and needles in the flesh (12).

Human nails contain about the same concentration of sulfur as human hair (3.3–3.5 mg/100 g fresh nails) and about half the concentration of zinc (10.8 mg/100 g dry weight) There is a trace of chromium in nails as in hair, but other components have not been measured. Nails are known to consist largely of keratinous proteins (complex polypeptides with disulfide cross linkages).[21] Neither these proteins nor trace minerals would have any known therapeutic effect on the four urinary problems, the chronic bloody diarrhea, swollen belly in infants (probably either parasitic infestation or malnutrition), unexpelled placenta, eye infection in smallpox, or needles in the flesh. The treatment for the latter, which combines powdered fingernails with jujube flesh, would not irritate the skin enough to cause loosening of needles or splinters.

The remaining five entities might be partially alleviated by the prescriptions, but in nonspecific ways. As for foreign bodies in the eye (13 and 14), the protein coagulum formed by combining powdered nails with either saliva (13) or rush pith (14) would become a sticky mass to which the foreign body would adhere and could be lifted out of the eye. Any other adhesive substance, especially when moistened with tears, could serve the same purpose. Powdered nails might have some limiting

[21] Altman and Dittner, *Biology Data Book*, s. v.

action in stopping epistaxis (18) by forming a proteinaceous coagulum over the bleeding point. Disorders due to wind agents entering open wounds (if this refers to infected open wounds) might be palliated by the direct application of the prescription (3), but the mixture taken by mouth would be ineffective even for palliation.

There is one lesion in this section, ulcerated hemorrhoids (11), where a certain degree of therapeutic specificity is possible, not for nails themselves, but for the additive, silkworm cocoons. The specificity applies only to cases of bleeding hemorrhoids, but most ulcerated hemorrhoids do bleed. It is known that cellulose, especially if oxidized, promotes clotting by a chemical reaction between the hemoglobin of blood and cellulosic acid. This might, then, be an example of the additive in a prescription having greater healing effect than the drug itself. For the pain of the ulcerated hemorrhoids, a poultice of cocoons, nails, hair, and honey might have a nonspecific palliative effect.

It is interesting that hair and nails were used as diuretics and as uterine myotonic agents for placental expulsion. Unless these appendages contain unknown substances that might act specifically to increase urine flow, there would be no rationale for their use. One might speculate that an induced mild infection is responsible in both cases for the contraction of the uterus which expels the placenta. The ritual act incorporated in the nail therapy (item 10) would be likely to induce in the patient a response conditioned by the expectations of her society.

III. Teeth 牙齒
1. Black pustules in the eruptive stage of smallpox 痘瘡倒靨. Three prescriptions are given; one also calls for the teeth of cats, pigs, dogs, and so on.

Human teeth are roasted carefully without physical decomposition and mixed with a little musk. Half a *ch'ien* (2 g) is taken with warm wine.
2. Nonsuppurating breast abscesses 乳癰未潰

Roast and grind human teeth, mix with butter fat, and spread on the abscess.

3. Various discharges from the ear 五般聤耳

For discharges of pus, blood, or water. Roast human teeth carefully to avoid physical decomposition, and powder with a little musk. Blow it into the ear. This is called "Buddha's-tooth powder."

4. Fistulary lymphadenitis and other draining skin ulcerations 漏瘡惡瘡. The former is consistently described in the medical classics as breaking out on the neck or near the armpits; this identification is due to Yü Yun-hsiu (*Y* 10–11; *C 34:* 179).

Dried fluid forming on the skin (? 乾水生肌). Take equal parts of the ash of human teeth, the ash of oily hair, and the ash of rooster's stomach lining mucosa 雄雞內金 and powder them.[22] Add a small amount of musk and crude calomel 輕粉 (Hg_2Cl_2). Make into a paste with oil and apply.

5. (We omit a fifth disorder 陰疽不發 because of our inability to identify it with reasonable confidence).

COMMENTARY

The four entities treated with human teeth are all external lesions due to infectious agents, although the pustules of smallpox are the dermatologic manifestation of an internal disease caused by a virus. Two of the prescriptions employ musk as an additive. Musk is the dried secretion from a special follicle of the musk deer (*Moschus moschiferus*), which inhabits mountainous regions of China's northern provinces. The secretion musk has a very strong sweetish odor and is used as a fixative in perfumes. This property is due to a viscid, volatile oil containing a ketone with the odor of musk; musk also contains fat, resin, cholesterin, proteins, and ammonium and calcium salts. Musk is said to be a nervine and antispasmodic when taken internally, but the biochemical explanation for this is obscure.[23]

[22] For an identification of the latter substance, whose name is literally "inner gold of rooster," see *L 48:* 78.

[23] See E. N. Gathercoal and E. H. Wirth, *Pharmacognosy*, by E. P. Claus (3rd ed., Philadelphia, 1956), pp. 317–318.

Enamel, dentin, and cementum of teeth are all calcified tissues containing both organic and inorganic matter. The principal inorganic salt, hydroxyapatite, is a complex salt of calcium, phosphate, and hydroxyl ions. The average composition of human enamel and dentin suggests no rational basis for the use of teeth in treatment of the lesions listed.[24] All of these lesions are due to infectious agents, and there are no bacteriostatic properties in powdered teeth nor in any of the admixtures used, with the possible exception of calomel.

IV. Milk 乳汁

Li Shih-chen is summing up a millennial tradition when he says, "It seems that milk is transformed from yin blood. It is formed in the spleen and stomach, and taken up by the *jen* and *ch'ung* circulation branches 生于脾胃, 攝于衝任. Before a child is conceived [the fluid] descends as menstrual blood; after conception it stays in place to nourish the foetus. After the child is born its scarlet color turns to white and it ascends as milk." He specifies that milk used for medicinal purposes come from a healthy woman whose first baby was male, and that the milk be white and thick. "Milk has no determinate physiological nature 性. If the person is harmonious and mild and her diet bland, her milk is bound to be balanced [i.e., neither cooling nor heating]. If she is hot-tempered, and given to drinking wine and eating strongly flavored food, or has a Fire disorder, her milk is bound to be heating. Milk should always be drunk hot. Some evaporate it to make a powder that is added to medicines; this is particularly good" (*L 52*: 96–97).

1. Debilitation due to *ch'i* exhaustion 虛損勞瘵. One of a large class of diseases characterized by a deficiency of homeostatic *ch'i;* in men, the symptoms are a floating but strong pulse, debilitation of the extremities, and coldness of the sexual organs with involuntary exudation of semen (nocturnal emission). It is acute in spring and summer and remits in fall and winter (*C 3*: 17b).

Virtuous Birth Elixir requires three wine-cups of milk from a

24 W. S. Spector, ed., *Handbook of Biological Data* (Philadelphia, 1956), s. v.

woman free of disease.[25] A porcelain saucer is set in the sun until quite hot. Then the milk is put into it, followed by a small amount of musk powder and 2 *fen* (19 g) of powdered elecampane root 木香 (*Inula helenium,* L., and related species). They are mixed until homogeneous and taken, followed by a small liquor-cup full of strong tea. This will overcome the yang. The next day take Connected Destiny Elixir. For Connected Destiny Elixir, take three wine-cups of milk. Expose a saucer to sunlight and put the milk into it as before. Add a powdered human placenta, mix, and take. After taking, the patient's face and knees will turn scarlet, and he will be sleepy, as if drunk. Give him only a small amount of white congee for nourishment.

Remarks: There is no discrete disease entity in modern medicine that combines weakness (of the extremities), a strong pulse, and nocturnal emission. The latter symptoms usually occur in postpubertal teen-age boys who have not yet established a normal level of sexual activity; in such persons they would not likely be associated with weakness. "Involuntary exudation of semen" may also have meant incontinence, or involuntary dribbling of urine, a frequent symptom of prostatic hypertrophy in older men. If the latter were true, the limb weakness, coldness of sexual organs, and strong pulse could have been associated symptoms of anemia, either due to poor eating habits in senescent males or to a variety of other afflictions of old age that have anemia as a clinical sign. We assume the presence of the latter symptom complex in this instance. Regardless of the etiology of the anemia, human milk would be beneficial because it is a complete food, containing a balanced mixture of high-quality proteins, and all the essential vitamins and minerals. Statistically, most anemias are the result of iron deficiency, and milk is rather low in iron. However, "Connected Destiny Elixir," taken

[25] As often happens, neither of the elixir names in this recipe can be translated with confidence, since one can only guess at their syntax and at the precise shades of meaning of their component graphs. This one, for instance, might be rendered with equal plausibility "Virtue-engendering Elixir." Such problems can be solved only when the graphs are seen together in another context.

on the day following the milk, contains powdered human placenta, which has large quantities of iron. This is a rather unusual example, then, of the two parts of a prescription supplementing each other; neither part by itself would be a complete treatment. The additives musk (described earlier), elecampane root (which is rich in the diuretic inulin), and congee add nothing to the potency of the prescription.

2. Wind diseases consequent upon *ch'i* exhaustion 虛損風疾. In Chinese medicine the term "wind disease" has three senses, which are often confused. In the broadest meaning, it refers to the six pathological agents that can affect the body at any point in the cycle of the seasons: wind, cold, heat, damp, aridity, and fire. The normal manifestations of these agents play basic roles in normal growth processes, and their abnormal manifestations cause disease. In a more specific sense "wind disease" refers to syndromes due to the first of this group, the wind instrumentality, alone. Many diseases due to invasion of the body by wind factors are nervous, paralytic, arthritic, or rheumatic in nature. The third sense appears in writers who use the word "wind" to apply to diseases that resemble those caused by external wind factors but whose etiology is internal, arising in the main from desire and weariness of mind.[26] Since exhaustion or weakness of the blood or *ch'i* is usually the predisposing condition for the invasion of wind factors, it is probable that Li Shih-chen's entity belongs to the second sense. In any case this entity is too broadly defined to be identified with any Western syndrome (*C 4*: 27; *K* 122).

Connected Destiny Elixir is used to treat exhaustion disorders in men or women in whom *ch'i* and blood are debilitated and phlegm and internal fire are ascending.[27] It is also effective for voicelessness due to wind disorders, paralysis of one side of

[26] For the internal factor Lu and Needham have adopted the term "blast," in analogy with van Helmont's *blas*. "Medicine and Chinese Culture," in *Clerks and Craftsmen,* p. 271.

[27] Note that this recipe for Connected Destiny Elixir differs from that in the preceding prescription. The discrepancy is not out of the ordinary, seeing that Li takes the two from different sources.

the body with limitation of mobility [literally, "slow movement"] in the other [=stroke? 癱左緩右], aching hands and feet, difficulty in moving about, and defective appetite (? 飲食少進). Take two wine-cups of human milk—that which is aromatic, sweet, and white is best—and mix with one wine-cup of good pear juice until homogeneous. Bring just to a rolling boil 頓滾 in a silver or stone vessel. Take every day before dawn [literally, "at the fifth night watch"]. It is able to reduce phlegm, remedy [ch'i] exhaustion, cause the formation of blood, and lengthen the life span. This is using man to fortify man—unsurpassably wonderful.

Remarks: As was pointed out above, this group of diseases cannot be identified with any specific entities known to modern medicine. If the blood is debilitated (anemia?), milk would not be beneficial because of its low iron content. The addition of pear juice to this recipe for "Connected Destiny Elixir" is interesting, because native pear juice is rich in vitamin C, which assists in the absorption of iron from the stomach and is useful in the treatment of secondary iron-deficiency anemias and the anemia of scurvy. However, vitamin C is destroyed (oxidized) by boiling, as are most of the vitamins in milk. Consequently, the method of preparation of this elixir deprives it of any value as a nonspecific tonic.

3. Voicelessness due to wind disorder 中風不語. The spleen section of the *ch'i* circulatory system originates at the end of the great toe of the right foot, and proceeds upward from the spleen, joining the base of the tongue and ending under the tongue. A separate branch connects the spleen and the heart. The attack of a pathological wind agent upon the spleen and the heart can be transmitted along the circulation to render the tongue rigid so that speech is impossible (*C 1*: 1; *T 19*: 528; clearly labeled diagram in *K* 85).

Base of the tongue rigid. Mix and grind together [*sic*] five *ko* (535 ml) each of three-year-old soy sauce and human milk and force the liquid part through unprocessed cloth. Give the patient small amounts at no set interval; after a long period he will be able to speak.

4. Sudden voicelessness 卒不得語. This disorder may overlap the preceding or following, since the three are cited from different sources.

Mix and take half a *ko* (54 ml) of human milk with half a *sheng* (535 ml) of fine wine.

5. Loss of voice 失音不語. Ch'ao's article on this disorder is worth quoting for its account of the basis of speech: "The windpipe is what *ch'i* ascends and descends through. The epiglottis 會厭 is the door of the voice.[28] The tongue is the mechanism of speech. The lips are the fans of speech. When a cold agent belonging to the wind category occupies the area of the epiglottis [= the larynx?], one is suddenly voiceless.... (*C 1*: 1). This description is too vague to be identified specifically with laryngitis.

Two *ko* (214 ml) each of human milk and bamboo sap [collected from heated short lengths of bamboo] 竹瀝 are taken warm.

6. Amenorrhea 月經不通

Drink three *ko* (321 ml) of human milk daily.

7. Eye feverish with red swellings 眼熱赤腫 (= trachoma?). Since the eye and the liver functionally correspond, a hot *ch'i* (energy of specific quality or configuration) from the liver can also attack the eye. If the body is simultaneously exposed to a cold agent in its ambiance, the antagonism of the two *ch'i* can give rise to a swelling just inside the eyelid, from the size of an apricot pit to that of a jujube (*C 28*: 147, s.v. 目風腫).

Half a *ko* (54 ml) of human milk and ten old copper coins are rubbed together in a copper vessel until there is a change in color, and cooked until the liquid is thick. Store in a narrow-necked bottle and drop in the eye several times a day. Some

---

[28] There is some confusion in the Chinese conception of the throat's structure, but the term I translate "epiglottis" generally denotes the valve that was believed to open when sound was produced and to close when food was swallowed.

Ch'ao's account is based on a passage in one of the two constituent books of the Yellow Emperor's Inner Classic, namely the *Ling shu ching* 靈樞經 (The divine pivot; ed. Liu Heng-ju 劉衡如; Peking: People's Hygiene Press, 1964), sec. 69, pp. 216–217.

practitioners soak coptis root 黃連 (*Coptis teeta,* Wallich or
Coptis chinensis, Franchet) in human milk. The milk is steamed
until hot and the eye bathed with it.

Remarks: This is perhaps an example of synergism between
the human drug, milk, and an additive, copper. It is possible
that the sulfur of milk and the copper react with the oxygen of
air to form small amounts of copper sulfate, which would
produce a green color change in the mixture. Before the emergence of antibiotics for the treatment of eye infections, copper
sulfate, either in stick form or mixed with glycerin, was used,
especially in the treatment of the eyelid inflammation and
hypertrophy of trachoma, a viral disease of the eyes prevalent
in the Orient.[29] Copper sulfate also has a mild antibacterial
effect in eye infections other than trachoma (including styes),
but its use is attended with some hazard because of the irritant
effect of copper sulfate upon the conjunctiva.

Coptis root contains the alkaloid coptine, formerly used in
medicine as a tonic and an astringent, especially for sores in the
mouth. It would not be useful for trachoma or any other type
of eye infection.

8. Urine retention in newborn babies 初生不尿. A general term
for congenital obstructions of the urinary tract. Ch'ao does not
list this disorder, but only "difficult urination in infants 小兒小
便不通利" (*C 49:* 262). Another source claims that this preparation is good for an infant's inability to suckle (*T 82:* 2610).

Four *ko* (428 ml) of human milk and one *ts'un* (3 cm) of the
white part of scallion are brought to a rolling boil and divided
into four doses, which will be effective 即利.[30]

9. Vomiting of milk in newborn babies 初生吐乳

Two *ko* of human milk, a bit of splint from bamboo matting
簟篿蔑, and a piece of salt the size of a couple of millet grains
are heated together to the boil. A piece of "cow bezoar 牛黃"
(biliary and other calculi from water buffalo) the size of a grain

---

[29] See C. A. Perera, ed., *May's Manual of the Diseases of the Eye* (20th ed.,
Baltimore, 1949), p. 127.
[30] *Or* after which the patient will urinate freely.

of rice or so is added and the medicine is given to the patient.[31]

10. Abscesses that will not burst 癰膿不出. This title is slightly ambiguous, but in Li's source the recipe is entitled "prescription for bursting a purulent abscess" (S 22: 20b–21a).

Human milk is mixed with wheat flour and applied externally. By dawn all the pus will come out. Manipulation is to be avoided.

11. Sores on the calf 臁脛生瘡

Equal parts of human milk and tung oil are mixed until homogeneous and brushed on with a goose feather. Miraculously effective.

Remarks: Here is an example where the additive is more effective as a healing agent than the drug itself. Oil extracted from the seeds of the tung tree (any of several Asian trees of the genus *Aleurites*) is used as a drying agent, and therefore could be helpful as a desiccant in the treatment of sores on the calf. Milk itself would be ineffective here.

12. Poisoning due to beef from an animal which has eaten a snake 噉蛇牛毒

If a cow has eaten a snake, its hairs will point toward its rear, and its meat will be fatally poisonous. But if the patient drink a *sheng* (1074 ml) of human milk he will recover immediately.

13. Poisoning from beef or horsemeat 中牛馬毒

Drinking human milk is excellent.

14. Invasion of the ear by insects 百蟲入耳

Drip human milk into the ear and they will come out.

COMMENTARY

Of the fourteen disorders treated with human breast milk, we have predicted possible therapeutic benefit in three instances, two of which may have been helped because of the additives rather than by the drug itself. The anemia of elderly men, regardless of etiology, would benefit from the combination of the two elixirs described under item 1. The other two instances

---

[31] Substantially the same prescription is also given by Li Shih-chen s. v. "matting 薦" (L 38: 46). For further references on "cow bezoar," see N. Sivin, *Chinese Alchemy*, p. 284.

of possible salutary results from the prescriptions are both due to the additives in the prescriptions, and have been discussed (items 7 and 11, remarks). None of the other eleven disorders could have been healed by known properties of human milk or by the additives described in individual cases, except in the case of item 14 (insects in ear), where milk would float out the insect, as would any other liquid.

Since human breast milk has a complete protein content of high biological value, and contains most of the vitamins and minerals as well as carbohydrate and fat, it is a complete food.[32] Li Shih-chen apparently did not recognize this fact, however, because he did not prescribe the use of milk for disorders due to malnutrition or inanition except in one instance (item 1).

V. Blood

A. MENSTRUAL BLOOD OF A MARRIED WOMAN 婦人月水

Li Shih-chen says that the menstrual rhythm corresponds (in categorical terms) to that of the monthly cycle of spring and neap tides, also controlled by the phases of the moon.[33] He notes many variations due to excess or defect of blood— discharge once in three months, once a year, or never; discharge during pregnancy; and periodic discharge through mouth, nose, eyes, or ears [!]. He decisively rejects the alchemical use of the first menstrual flow of virgins in elixirs and other preparations, but strongly supports the taboo against sexual intercourse with menstruating women on the ground that their blood, because it is unclean, injures the man's yang, and causes disease. The prescriptions given below use the menstrual cloth 月經衣, which served the purpose of the modern sanitary napkin, as well as the blood itself.

[32] Altman and Dittner, *Biology Data Book* s. v.
[33] See Needham, *Science and Civilisation in China* (7 vols. projected, Cambridge, England, 1954–), III, 484–494.

1. Recurrence of a hot disease due to fatigue 熱病勞復. This is one of a broad subclass of the *shang han* febrile diseases, the members of which occur in the summer (*C 9:* 60).

When the husband has recovered from a disease due to hot factors, and there is a recurrence after sexual intercourse, suddenly his testicles contract into his belly and his intestines ache so that he feels about to die.[34] Roast the wife's scarlet menstrual cloth, powder it, and take one square-inch-spatula-full with boiled water, whereupon the symptoms will subside.

2. Yellow jaundice due to fatigue in women 女勞黃疸

With short breath and muffled voice 聲沈. Take the woman's menstrual blood, mix it with the menstrual cloth, and roast to ash. A square-inch-spatula-full is taken with wine twice daily; the disease will remit after three days.

3. Acute sudden delirium 霍亂困篤. The term "*huo luan* 霍亂," which in modern Chinese is the official translation for "cholera," covers a number of closely related syndromes caused by imbalance between warm and cool factors in the body, leading to conflict between the yin and yang *ch'i* in the stomach and intestines. The most common precipitating factor is eating large quantities of food inappropriate to the time of year. The symptoms—vomiting and diarrhea with griping pains in the stomach and heart, delirium preceding death—include those common to simple and malignant cholera, and to some extent of food poisoning (*C 22:* 121; *S 20:* 132–142).

Mix the menstrual cloth of a virgin girl with her blood and roast to ash. A square-inch-spatula-full is taken with wine. This is used when other prescriptions fail.

4. Convulsions in children due to fright 小兒驚癇. This disease belongs to a group of fright disorders that includes epilepsy (*C 45:* 241).[35]

[34] Note the similarity of the symptoms to those described by P. M. Yap in "Koro: A Culture-bound Depersonalization Syndrome," *British Journal of Psychiatry,* 1965, *111:* 43–50.
[35] Cf. item I.4 in this chapter. The *Mei shih fang* 梅氏方 (Sui?) gives essentially the same treatment for rabies (*H 15:* 7a).

With fever. Take the monthly blood 月候血 and mix it with indigo.[36] Take one *ch'ien* (4 g) in water. The attack will cease as soon as the medicine enters the child's mouth. Use more or less according to the size of the child.

5. To keep one's wife from being jealous 令婦不妬. This "prescription" does not originate in the medical literature, but in the Record of Wide Knowledge (*Po wu chih* 博物志 of Chang Hua 張華, ca. A.D. 290), a collection of marvels. The original is long lost, and this magical recipe is not found in the current reconstituted version.

Wrap a toad in the cloth the wife uses to absorb her menses 婦人月水布 and bury it five *ts'un* (15 cm) deep, one *ch'ih* (31cm) in front of the privy.

6. Carbuncles on the back 癰疽發背. It is difficult to link Ch'ao Yuan-fang's account of the etiology and character of this group of disorders—differentiated by associated symptoms such as diarrhea and constipation—with Western syndromes, although one might speculate that there is at least some overlap between his "carbuncles on the back with thirst 癰發背渴" and diabetes. The cold abscess form can be mortal; in fact the alchemist-physician Sun Ssu-mo, writing in the late seventh century, notes that it is usually due to ingestion of minerals, and commonly fatal within ten days. He continues, "there are those who get carbuncles on their backs without ever in their lives taking mineral preparations; that is because among their ancestors there was someone who took them" (*C 33:* 175–177; *S 22:* 26a).

For all ulcerous toxicity. Take equal quantities of swallow's nest 胡燕窠, fine earth dug by field mice 土鼠坌土, white bark of elm 榆白皮, and dried trichosanthis stem 栝樓根 (*Trichosanthes kirilowii*, Maximowicz), and powder them together. Wash a menstrual cloth with water, take the fluid, and mix it with the ingredients [to a paste]. Apply it to the abscesses and replace it as it dries. For burst abscesses, seal them in from all

[36] This may very well have been a native Chinese vegetable blue. See Sung Ying-hsing, *T'ien-kung k'ai-wu. Chinese Technology in the Seventeenth Century* (tr. E-tu Zen Sun and Shiou-chuan Sun; University Park, Pa., and London, 1966), pp. 75–76.

sides. Relief in five days.

7. Sores on the penis 男子陰瘡

Due to sexual intercourse in violation of the menstruation taboo; soft, easily bursting sores 潰爛 on the penis. On a piece of tile, roast to ash the cloth used by a virgin to absorb her menses 室女血衲. Grind the ash to powder, make it into a paste with sesame oil, and apply.

8. Arrow poisoning 解藥箭毒. This treatment also comes from Chang Hua's Record of Wide Knowledge.

The barbarians of Chiao-chou [mainly modern North Vietnam] make a poison from scorched copper [implements] 焦銅 and put it on their arrowheads. When someone is hit the wound suppurates 沸爛, and in a short time the bone degenerates. The only antidote is to drink menstrual blood and excremental fluid.

According to the current version of Li's source, the copper implements, whose poisonous virtue is determined by listening to their sound when struck, are made *into* arrowheads, and an additional poison smeared on their points. Nothing is said about the bone's degeneration. [37]

9. Arrowhead lodged in the abdomen 箭鏃入腹. This remedy comes from the Prescriptions Worth a Thousand of Sun Ssu-mo. In order to illustrate the freedom (or carelessness) with which Li Shih-chen often adapts his sources, I translate the source first (*S 25:* 36a):

From Sun: Prescription for treating a soldier who has been wounded by an arrow or crossbow bolt which will not come out. In some cases there is an accumulation of blood. Take the cloth used by a woman to absorb her menses, roast to ash, powder, and take with wine.

From Li: Arrowhead lodged in the abdomen. In some cases there is an accumulation of blood in the flesh. Take a woman's menstrual cloth, roast to ash, and take a square-inch-spatula-full with wine.

10. Horse's blood entering a sore, or a wound [from a broken

[37] *Po wu chih (Pai tzu ch'üan shu* 掃葉山房百子全書 ed.), *2:* 2b. Excremental fluid is the liquid that gradually forms in an open privy.

bone end] piercing the skin while skinning a horse 馬血入瘡, 剝馬刺傷. According to the fuller description in the Pharmacopoeia of 1249, the wound leads to fatal poisoning (*H 15:* 6a).

Applying menstrual blood is miraculously effective.

11. Sore wounds from tigers and wolves 虎狼傷瘡

Roast a menstrual band to powder and take a square-inch-spatula-full with wine three times a day.

B.   HUMAN BLOOD IN GENERAL 人血

1. Prolonged hematemesis (vomiting of blood) 吐血不止

Use the [dried?] clots of blood that have been vomited; fry them black and powder them. Take three *fen* (28 g) per dose, mixing it with broth of ophiopogon stem 麥虋冬湯 (*Ophiopogon japonicus,* [Thunb.] Ker-Gawler) before taking. It seems that because the blood does not return to its origin, it accumulates and reverses its course upwards. If blood is used to guide the blood, it returns to its origin and the bleeding stops.

2. Prolonged epistaxis (nosebleed) 衄血不止. Two similar treatments, the first of which is as follows.

The Imperial Compendium of Prescriptions of 1111/1117 聖濟總錄 uses a sheet of white paper, with which the blood is received until it is saturated. Then the paper is burnt to ashes over a lamp to make one dose, which is taken with freshly drawn water. Do not let the patient know.

3. Internal hemorrhage from wounds 金瘡內漏

Take blood exuded from the wound, mix with water, and drink.

4. Postpartum nausea with abnormal blood loss 產乳血運. This disorder, which can be fatal if uninterrupted, is characterized by too great or too little blood loss during delivery. In the former case the blood is weakened and overbalanced by the *ch'i;* in the latter, the *ch'i* reverses its course and the blood, following it, stops up the heart. These irregularities are said to be generally caused by violation of taboos concerning what directions pregnant women should face (as given by the Five-Phases theory) at the time of delivery. Here is how Ch'ao Yuan-fang puts it (*C 43:* 230b): "When a woman violates the

prohibitions concerning the directions toward which one faces when sitting or lying down during labor, nausea will usually result. Thus the downflow of blood will be either excessive or deficient. For this reason, the direction she faces when sitting or lying down in labor should, appropriately to the season, observe the prohibitions according to the Five Phases. If these are violated, one will generally call down calamity." It is all too probable that a woman in labor, upon realizing that she has violated a taboo, would experience nausea, and that it would affect her bleeding. Only a foolish doctor practicing among traditional Chinese would deny that the best short-term prevention is careful observance of the taboos (see also *W* 557–558; not in *T 80:* 2522–2529).

Take concentrated vinegar and mix it with blood collected at the time of delivery [and coagulated], the size of a jujube, and give it to the patient to ingest.

5. Scarlet birthmarks in infants 小兒赤疵

Stick a needle in the arch of the father's foot 腳中. Take some blood and rub it on the birthmark, which will thereupon fall off 落.[38]

6. Warts in infants 小兒疣目. This treatment is condensed to the point of ambiguity, but reference to Li Shih-chen's source makes the meaning clear (*S 5B:* 18a–18b).

With a needle break the surface of the wart all around. Take pus from someone with a sore and rub it on. Avoid contact with water for three months, and the wart will fester and fall off.

COMMENTARY

Two types of human blood were used in treatment: menstrual blood of a married woman and ordinary blood. These categories will be discussed separately.

MENSTRUAL BLOOD OF A MARRIED WOMAN

Scientifically speaking, none of the disorders treated with menstrual blood would be alleviated by its use. Menstrual blood, whether from a married or an unmarried woman, consists of blood, mucus, and fragments of endometrium

[38] The source, *S 5B:* 18a, reads "go down 消."

(uterine tissue). It contains fewer red blood cells and more white blood cells, especially lymphocytes, than does the circulating blood. Menstrual blood also contains an unidentified toxic substance called menotoxin, and fibrinolytic enzymes. The presence of the latter enzymes causes menstrual blood to appear incoagulable; the blood actually does clot inside the uterus, but the presence of fibrinolysins causes the clots to liquefy after intrauterine coagulation. Fresh human menstrual blood also contains proteases capable of dissolving clots of peripheral blood.[39]

Because of the cyclic nature of the menses, whose rhythm was believed to correspond to the phases of the moon, one might expect that the disorders treated by menstrual blood would be cyclic, or periodic, disorders. This is not true of Li Shih-chen's selection, with the possible exception of item 1 (recurrence of "hot" disease due to fatigue), which could correspond roughly to malaria, a recurrent febrile disease. None of the other ten ailments is recurrent, and no other common causal factor is apparent.

On the other hand it would be inadvisable, in view of the strength of the taboo against intercourse with a menstruating woman, to ignore possible ritual and magical origins for these treatments. Item 5 is obviously magical, depending upon the notion that ritual operations upon the menstrual cloth will bind the woman's emotions. The disorder in item 7 is accounted for explicitly by violation of the menstrual taboo. The idea of sympathy, which we have seen Li Shih-chen invoking overtly but in an abstract form with reference to blood in general ("if blood is used to guide the blood . . .") probably plays at least a subsidiary role in therapy with menstrual blood, since items 8 through 11 involve blood and bleeding, and 8 is a classical example of what in the West was called *Dreckapothek*.

HUMAN BLOOD IN GENERAL

We can be more positive, however, about therapeutic benefit from the use of human blood in general. Benefit is likely to have

---

[39] See N. J. Eastman, *Williams' Obstetrics* (10th ed.; New York, 1950), pp. 101–102.

occurred in four of the five conditions for which it was used. We omit item 6 (warts in infants) from the discussion because in this case blood was not used specifically as a therapeutic agent. Indeed, the application of pus to warts would only aggravate the condition by causing secondary infection. The therapeutic benefits of blood would come chiefly from its high iron content and its good protein quality. Although blood contains all the amino acids and vitamins, the low concentrations of these substances would make blood a less useful source of them than most foods of animal origin.[40]

Human blood, because of its high iron content, would be potentially useful for any type of blood loss. Blood loss results in iron-deficiency anemia and is treated with iron salts in order to rebuild the body stores of hemoglobin, the oxygen-transporting red pigment of erythrocytes. Hemoglobin is a complex of heme, an iron-protoporphyrin chelation compound, and globin, a protein. The normal human body manufactures heme (and globin), but must have an exogenous supply of iron to make the final product, hemoglobin. If, for one reason or other, iron either is not absorbed or is lost externally, hemoglobin is not produced in adequate amounts, and anemia results. The absorption of iron from the stomach and intestines is assisted by an acidic environment; hence the vinegar (acetic acid) given with blood in the treatment of postpartum nausea due to abnormal blood loss might facilitate iron absorption. In the cases of hematemesis and epistaxis, absorbable iron would still be present in the administered blood, even though the blood had been treated by frying and ashing. Internal hemorrhage from wounds is treated with wound blood mixed with water; in this case, ferrous iron is available for absorption after it has been separated from heme during the process of digestion. As in the case of bone in the treatment of bone disorders, blood was probably used in these instances to treat disorders (that is, external loss) of human blood.

The general principle of restoring lost or faulty tissues with

[40] Altman and Dittner, *Biology Data Book*, s. v.

their own kind may be grossly simplistic from the physiological point of view, but from the standpoint of magic it is merely an obvious application of the law of sympathy, which asserts that like affects like. We can see this law transformed to bring it into the highly rational structure of theoretical medicine in Li Shih-chen's statement (item 1, also implied in item 4) about using blood to guide internal blood back to its origin. The pharmacological efficacy of blood in items 1–4 should not obscure the fact that in all four blood is being used to control blood. If the empirical factor helped secure the retention of these remedies, it is equally likely that the ritual factor is responsible for their origin in folk medicine. It would be foolish to close our eyes to the possibility that magic and empirical science play inseparable roles in early culture merely because we have generally succeeded since the Scientific Revolution in severing their connections, and because we prefer the tidiness of either/or questions. Item 5, which does not make clinical sense, can be pondered, at least, as a more complex example of sympathetic magic, the connection being that between the color of the birthmark and that of blood, with an overtone of the special efficacy of the infant's father.

Item 4 provides an invaluable concrete example of how the line between folk medicine and theoretical medicine is crossed. The directional taboos responsible for abnormal post-partum bleeding are explained not in terms of the vital and spirit-filled world of village belief in which taboos grew up, but by re-course to the classical tradition's rational (if anything, over-rational) Five-Phases theory. This transposition could as well have been made by an illiterate village practitioner as by Ch'ao Yuan-fang. The currency of the Five-Phases concept in China reminds us that while the experience of tens of millions of patients was being distilled out of folk medicine, there was a corresponding downward movement, however limited, of theoretical concepts originally developed by scientists of the high tradition.

VI. Semen 人精

Li Shih-chen speaks of female seminal fluid as well as male

semen; the former, which has no other common name in China, is almost certainly the vaginal lubricant secreted during the foreplay or the sexual act.[41] At the same time, the Chinese definitely believed that at the moment of orgasm the woman released an essence which was the yin counterpart of the male semen. Whether this was considered material (and *confused* with the lubricant fluid) is generally not quite clear.[42] In the prescriptions given here it is used only in the small amounts necessary to rub on the skin. In Li's time black magicians 邪術家 were mixing it with male semen to make a potion called "lead and mercury" for use in sexual alchemy.[43] He believed that a sixteen-year-old male possesses over a liter and a half of semen, which can increase with age and cultivation to a maximum of nearly five liters.

The word "*ching* 精" has both a broad and a narrow meaning in Chinese medicine; in the former sense it is sometimes rendered "seminal essence." It is the finest essence of ingested nutriment, fermented in the stomach, assimilated by each of the internal organs, and stored in the kidneys, from whence it is recycled to the other organs as needed for their sustenance. The semen which takes part in the reproductive process is the essence proper to the kidneys themselves, imbued with yang *ch'i*. This reproductive essence is stored separately in an organ called the "gate of destiny 命門," variously identified in the

[41] Cf. Robert H. van Gulik, *Sexual Life in Ancient China. A Preliminary Survey of Chinese Sex and Society from ca. 1500 B.C. till 1644 A.D.* (Leiden, 1961), pp. 45–46.

[42] For instance, in a passage on sexual hygiene in the compendium *Ishin hō* 醫心方 (A.D. 982; Peking, 1955), *28:* 639a, the last of the ten motions that mark the stages of coition is given as "that the fluid in the sexual organ is slippery, as a result of the semen's already having been ejaculated." This would almost certainly refer to the man's semen, although there is also a possibility that it is being applied to a female seminal fluid that would affect the slipperiness of the already lubricated vagina. The version of Ishihara and Levy, which tends toward Chinoiserie, reads here (p. 42) "the essence having already leaked out."

[43] There is much data on the extravagances of late sexual alchemy in the Pharmacopoeia. One can find interesting background information in Liu Ts'un-yan, "Lu Hsi-hsing and his Commentaries on the *Ts'an t'ung ch'i*," *The Tsing Hua Journal of Chinese Studies*, 1968, n. s., *7.* 1: 71–98; but this article is full of technical errors and mistranslations.

medical literature as the right kidney or the space between the kidneys.

1. Facial lentigo 面上䵟子 (*Y* 178–179)

Human semen is mixed with the white part of hawk excrement 鷹屎白 and applied.[44] Recovery in a few days.

2. Pink growth on the face or body 身面粉瘤. In modern vernacular this is a term for cancerous tumors, but in classical medicine the disorder was one of a class of painless, soft, benign large growths (*C 31:* 163; *Y* 242; cf. *W* 511). There is no point at present in even suggesting an identification with any specific entity in modern medicine.

A *ko* (107 ml) of human semen is packed into a green bamboo tube and roasted over a fire. Catch the liquid in a receiving vessel and seal it tightly. Apply it to the growth repeatedly, ceasing when it has taken effect.

3. Ulcerated lymphadenitis 瘰癧腫毒. One of the nine major types of *lou* 瘻, which refers mainly to swellings of the lymph glands (see above, IV, 4). According to Ch'ao Yüan-fang, *lei-li* is caught when dampness sets into the neck, forming a connected series of two or three hard swellings that lead to occasional fever and, if not treated, become infected and develop fistulas (*C 34:* 179, 182; *Y* 119–120). This treatment is for the advanced form.

Apply female seminal fluid repeatedly.

4. Scalds and burns 湯火傷灼

To stop pain, and for easy healing without scars. The Easy Prescriptions (*Chou hou fang* 肘後方, attrib. Ko Hung 葛洪, early fourth century) uses human semen and the white part of hawk excrement, applying them daily. Prescriptions Worth a Thousand uses female seminal fluid, which is applied repeatedly.

[44] The Chinese term "*ying* 鷹" is as broad as "hawk," although of course the overlap is not complete; in ancient times the former often included falcons. Li Shih-chen's note (*L 49:* 17) that the medicinally superior type comes from the Gulf of Liaotung suggests an identification with *Accipiter gentilis*, Schvedowi (Menzbier), the most common species that winters there. See Cheng Tso-hsin 鄭作新, *Chung-kuo niao lei fen-pu mu-lu* 中國鳥類分部目錄 (Distribution list of Chinese birds), Part 1 (Peking, 1955), pp. 44–45.

COMMENTARY

As is also true in the cases of teeth and saliva, semen is used in the treatment of ailments that are external, at least symptomatically. It seems rather a waste to limit the use of semen, the "finest essence of ingested nutriment," to these superficial lesions. More consonant with general theoretical principles would seem to be the use of semen for diseases due to yang deficiency.

Human semen contains spermatozoa and the secretions of the seminal vesicles, prostate, Cowper's glands, and, probably, the urethral glands. Semen is highly buffered and is slightly alkaline, a quality that would give it a nonspecific soothing effect on burns.[45] However, when mixed with hawk excrement, containing large quantities of bacteria, the prescription would serve only to aggravate matters by leading to infection of the burned area if the skin were broken. There is no possible benefit of semen for facial lentigo (freckles or nevi) or for tumors of the skin, whether benign or malignant.

Ulcerated lymphadenitis—swollen, draining lymph nodes and lymphatic channels in the neck—probably was a description of scrofula, or tuberculous lymphadenitis. This disease would not be aided by the application of any of the secretions of the female genitourinary tract. The term "female seminal fluid" has no modern counterpart. Formerly it was thought that during sexual foreplay the Bartholin's glands on either side of the vagina produced a lubricant useful during intercourse. The mucoid substance is now known to be a secretion of the vaginal lining itself. As we have suggested earlier, the term "female seminal fluid" perhaps referred to this secretion.

VII. Saliva 口津唾

Li Shih-chen identifies four salivary glands, all sublingual. One pair condenses *ch'i* from the heart, and the other seminal essence stored in the kidneys. When the saliva is swallowed these

[45] N. S. Spector, ed., *Handbook of Biological Data* (Philadelphia, 1956) and W. F. Ganong, *Review of Medical Physiology* (3rd ed.; Los Altos, California, 1967), s.v.

essences recycle to "irrigate the internal organs and moisten the limbs"; Taoist adepts systematically swallowed saliva while performing their breathing exercises in order to make use of this rejuvenating function. As a natural corollary, spitting was avoided as wasteful of vitality. In order to preserve eyesight, Li recommends that every morning after rinsing the mouth and brushing the teeth, one lick his thumbnail and use the saliva to wash his eyeballs. Having someone lick one's eye is recommended for "cloud films 雲瞖," which is more likely a mucous "false membrane" than a cataract.

1. Infection and pain around the fingernail 代指腫痛 (= felon). According to Ch'ao, this syndrome begins with swelling and pain in the finger, but without blackening; no trauma is mentioned. The skin around the nail subsequently becomes purulent and the nail drops off (C 30: 161).

Mix saliva with pure sal ammoniac 白硇砂 (NH_4Cl), make into a dough with flour 搜麪, and form a bowl out of it. Fill the bowl completely with saliva, add a little sal ammoniac, and soak the finger in it. It will be relieved within a day.

2. Warts on the hands and feet 手足發疣. Ch'ao Yuan-fang distinguishes these *yu mu* 疣目, which occur "on the sides of the hands and feet" and would thus be verruca vulgaris, from "rats' nipples" (*shu ju* 鼠乳), which can occur anywhere on the body and thus include the other common types of wart. The denotation of the former term seems to have broadened in later writing as the latter term disappeared (C 31:163; Υ 9–10, 263).

Take white millet flour 白粱米粉 (long-panicled *Setaria italica*, L. or similar) and fry until brown in an iron kettle. Grind to powder and mix with spittle from a number of people. Spread it on the affected place a *ts'un* (3 cm) thick, and the warts will go down.

3. Underarm odor 腋下狐氣

One takes one's own saliva, rubs it on the armpits several times, and scrapes off the dirt with the fingernails. Then wash the hands several times with hot water. After this is done for ten days or so the condition will be cured.

4. Snakebite 毒蛇螫傷

Quickly wash away the blood with urine. Then take spittle from the mouth and apply it repeatedly.

COMMENTARY

Saliva was also used as a remedy for external symptoms and ailments. Its principal physiologic function is to moisten and lubricate the oral cavity, and to assist digestion by mixing with ingested food to moisten it thoroughly and by providing the enzyme amylase to digest starches. Saliva is very rich in lysozyme, a mucolytic enzyme, and this might account for Li's recommendation that it be used to remove "false membranes" in the eye (if they consist largely of mucus).[46] Lysozyme is concentrated in tears themselves. It is therefore possible that in eye diseases involving blockage or failure of secretion of the lachrymal (tear) glands, saliva might be substituted for tears to rid the eyes of thick secretions. It would be hazardous to use saliva for any kind of eye disease, however, because of its high bacterial content. Saliva might easily cause eye infection— especially in the absence of tears, which have some bacteriostatic properties. No benefit could be anticipated by using saliva to treat warts, snakebite, or felons. The therapy for underarm odor amounts merely to cleansing.

VIII. Bone

A.   IN GENERAL 人骨

Li Shih-Chen expresses moral reservations about using human bone in medicine, for Chinese believed that the bones deserved special care as the ultimate remains of the dead body. In the following passage Li not only accepts the truth of a most remarkable story from the Yu-yang Miscellany (*Yu-yang tsa tsu* 酉陽雜俎, ca. 860), a collection of preternatural anecdotes, but in doing so interprets as an application of the theory of *ch'i* resonance what we immediately recognize as a classical instance of magic performed on an extension of the self—a bone

[46] Altman and Dittner, *Biology Data Book*, s. v.

chip that has been removed from the body but still affects it:

The ancients considered it benevolent and virtuous to cover exposed bones and often received preternatural rewards for doing so. But now wonder-workers, whose hearts are set on profit and the satisfaction of craving, collect human bones to make medicines. Can [medicine,] the "art of benevolence 仁術," actually be concerned with this sort of thing? And dogs do not eat the bones of dogs; can it be permissible for men to eat the bones of men? The bleached bones of a father—only the blood of his natural son, drawn and sprinkled on them, will soak in. Further, the Yu-yang Miscellany says, "A man of Ching-chou broke his shinbone. Chang Ch'i-cheng 張七政 gave him a medicinal wine to drink [as an anesthetic], pierced his flesh, removed a chip of bone, and applied a salve.[47] The patient recovered, but three years later his leg was aching again. Chang said, 'That is because the bone I took out is cold.' They looked for it, and it was still under the bed. They washed it in an infusion, wrapped it in silk floss, and put it away, whereupon the pain stopped." Such is the responsiveness of *ch'i*; who says that dry bones are without consciousness 知? "The benevolent" [that is, doctors] should be aware of this.

1. Effects of corporal punishment 代杖[48]

Roasted human bones are powdered, and three *ch'ien* (11 g) taken with wine on an empty stomach. When one is beaten with a stick no swellings or sores will develop. If it is taken over a long period the skin will also thicken.

2. For setting broken bones 接骨

One *liang* (37 g) of roasted children's bone, two *ch'ien* (7 g) of mastic, and a square of festive red silk gauze 喜紅絹 are roasted to ash, powdered, mixed in hot wine and taken. First splint the bone with slats of paulownia wood. Immediately effective.

3. For sores on the calf 臁瘡

[47] This is apparently a courtesy name, and we have not identified the surgeon.

[48] In this section all the prescriptions are given two-character titles; elsewhere they all have four-character titles.

Roasted human bone that has been crushed is powdered and applied.

4. Internal and external traumas from falls or blows 折傷. Bone from a dead child, after roasting, and muskmelon seeds fried until dry, are taken with good wine. Pain stops extremely quickly.

B. THE BREGMA 天靈蓋

The special importance of the topmost part of the skull, where the coronal and sagittal sutures meet (the site of the anterior fontanelle in infants), is due to peculiarities of esoteric practices that, although particularly associated with Taoism, were too widespread to be thought of as belonging to a single religious system. First, in esoteric anatomy (interior cosmography would be a better term) the cranium enclosed the Nirvana 泥丸, the topmost of the three centers of concentration of energy in the body called the Fields of Cinnabar 丹田. The object of the various disciplines—breath control, ingestion of elixirs, and so on—was (in Li Shih-chen's words) "to return to the pure yang state [of the newborn child]. When the Sage Embryo is perfected, an opening is made at the top of the head for it to pass through, which is why this bone was given such names as Supernal Sentience Cover." This embryo, which was essentially one's self nurtured and perfected until its restored spiritual energies and vitality were no longer subject to the attrition of time, discarded the mundane body like a butterfly hatching from the chrysalis, and assumed its new physical existence as a Taoist immortal. The drilling of the skull in order to provide a passage is still part of the initiation of members of the esoteric Buddhist Mi-tsung 密宗 (Shingon) sect in Taiwan today. Symbolically this ritual act recreates the unfused fontanelle of the newborn infant. The first recipe given below is typical of Taoist medicine, with its incantation, magic, and taboos, although the influence of Taoism upon medicine was so great by the T'ang period that every sort of ritual and magical procedure was common in the most highly regarded medical books. It is cited from the Purple Court Immortality Formulary, a lost work belonging to the Mao shan Taoist tradition 上清紫庭仙方.[49]

1. Supernal Sentience Cover Powder 天靈蓋散

Expels debilitating worms. A piece of Supernal Sentience [that is, the part of the skull surrounding the bregma] two fingers wide, after washing in an infusion prepared by boiling sandalwood, is scorched over a flame fueled with butterfat. This incantation is recited seven times in one breath: "Divine Father Thunder, Sage Mother Lightning, if you meet a cadaver vector you must control it. Quickly, quickly, as ordered by the law."[50] Five betel nuts 尖檳榔 (*Areca catechu*, L., or possibly *Capparis masaikü*, Léveillé), two *fen* (19 g) of asafoetida 阿魏 (*Ferula foetida*, Regel and related species), three *fen* of musk 麝香, one *fen* of cinnabar from Ch'en-chou in Honan 辰砂, three *fen* of gum benzoin 安息香 (*Styrax benzoin*, Dryander and related species), and three *fen* of kansui root 甘遂 (*Euphorbia kansui*, Liou) are powdered and taken in doses of three *ch'ien* (11 g).

Put four *sheng* (4.3 l) of urine from a boy into a silver or stone vessel and boil together with fourteen stems each of onion 葱白 (*Allium cepa*, L.) and shallot 薤白 (*Allium bakeri*, Regel), two bunches of artemisia 青蒿 (*Artemisia apiacea*, Hance), two five-*ts'un*-long (15 cm) pieces each of peachtree branch and licorice stem 甘草 (*Glycyrrhiza uralensis*, Fischer), and two seven-*ts'un*-long pieces each of willow branch, mulberry branch 桑枝 (*Morus alba*, L.), and pomegranate branch, until the liquid is

---

[49] Since Li does not include this book in the list of sources at the beginning of the Pharmacopoeia, he is probably quoting at second hand.

The standard source on Taoist anatomical conceptions remains Henri Maspero, "Les procédés de 'nourrir le principe vital' dans la religion taoïste ancienne," *Journal asiatique*, 1937, *229*: 182–197. The information on the piercing of the skull in modern initiations was furnished by Dr. Kristofer Schipper, of the École Francaise d'Extrême-Orient, who has served as a Taoist priest in Tainan, Taiwan. On the Mao Shan tradition see Holmes H. Welch, "The Bellagio Conference on Taoism," *History of Religions*, 1969–1970, *9*: 119–120 et passim.

[50] For "cadaver vector disease" see prescription no. 2 in this section.

reduced to one *sheng*. Divide the product into two portions.[51] At the beginning of the last night watch mix it [with the powder] and take it. After one dose of the aforementioned medicines, [if?] the worms are not purged, have the patient walked ten *li* (4.5 km) 約人行十里 and give him another dose. Then at dawn [of the next day?] give him still another dose. Catch the purged worms, which will be of assorted sorts and shapes, and put them into a kettle of oil and fry them. Those worms whose mouths 觜 are dark blue, dark green, scarlet, yellow, or brown can be treated; those whose mouths are black or white cannot be treated. In any case one can also determine the danger of contagion.

Before preparing this medicine, it is necessary to ritually purify oneself in a sanctuary far from human habitation. Do not let sick people [*or* the patient?] smell the *ch'i* [= aroma] of the medicine. Do not allow chickens, dogs, cats, livestock, mourning sons, women, nor anyone who has been in contact with uncleanliness, to see it.[52] After the worms have been purged, feed the patient congee as a restorative. Several days afterward he will dream of people tearfully parting; that is a sign of efficacy.

2. Bone-steaming disease caused by debilitation 虛損骨蒸. This is a synonym for "cadaver vector disease 傳尸," which passes from one member of a family to another, sapping their strength and killing them in turn. The name comes from the patient's feeling of feverishness ("steaming") in the marrow of his bones. The syndrome as described in classic medical works corresponds on the whole to that of pulmonary tuberculosis (see Sivin, *Chinese Alchemy*, p. 298, for fuller description and references). One of two prescriptions is given below.

---

[51] The bark of each of these trees is widely used in Chinese medicine. Note that although the liquid is divided into two portions, directions are given for three doses. Li's habitual carelessness in quoting probably accounts for this discrepancy.

[52] We translate 物 as "anyone" to make sense of the text. The word sometimes means "supernatural beings" in Taoist writing, but that reading does not make sense here. Careless copying is likely.

From Prescriptions Worth a Thousand. Use a piece of Supernal Sentience Cover the size of a comb. Scorch it brown. Simmer it in five *sheng* (5.4 l) of water until the volume is reduced to two *sheng*. Divide it into three doses. A miraculous medicine capable of raising the dead.

3. Bone-steaming disease in infants 小兒骨蒸 (*T 88*: 2821). Here Li's prescription differs sufficiently from that in his source that the discrepancy may be explained by actual clinical experience:[53]

Wang: Supernal Sentience Cover: one. Smear with butterfat and scorch until brown. Coptis root: one-half *liang* (19 g). Discard the hair[-like rootlets]. The above medicines are pounded and put through a fine sieve to make a powder. Mix into congee and take half a *ch'ien* (2 g) three or four times a day.[54] Regulate the dose according to the size of the child.

Li: Body emaciated and mind troubled. Scorch Supernal Sentience Cover over a flame of butterfat. Grind to powder with equal parts of coptis root. Each dose is half a *ch'ien* (2 g), drunk with rice [congee] twice a day.

4. Autumnal intermittent fevers 諸瘧寒熱. The Chinese divided the *shang han* febrile diseases into broad groups according to the relation between time of exposure and time of onset, thus relating them to the concept of disease as a breakdown in the phasing of the rhythms of the human organism with the rhythms of the seasons. The *wen* 瘟 diseases, for instance, were due to contact with a pathological agent in winter, but manifested themselves in spring and early summer. The *nueh* 瘧 diseases, the subject of this prescription, made themselves felt in autumn following exposure in summer. Like other members of the larger group of *shang han* febrile diseases, *nueh* includes malaria, typhoid, and other illnesses with similar symptoms (Sivin, *Chinese Alchemy*, pp. 302, 304).

Supernal Sentience Cover is roasted and ground to powder.

---

[53] Cf. the example given earlier (p. 237).
[54] There is an apparent textual error in the phrase 以粥飲調下; we make sense of it by interchanging the third and fourth characters.

One *tzu* 一字 (1 g) is taken with water.[55] Effective.[56]

5. Inability to eat caused by *ch'i* condensation in the diaphragm region 膈氣不食. There are five basic disorders in which emotional, climatic, or somatic factors cause a yin-yang imbalance manifested as a stoppage of the diaphragm (which the circulation system traverses), and consequent interference with the ability to swallow or to digest. A number of digestive dysfunctions are cited as symptoms, but we can suggest no identification with modern pathological entities (*T 50:* 1536).

Seven Supernal Sentience Covers are packed alternately with layers of forty-nine black soya beans (*Glycine max*, Merrill) each, sealed and made to rise and fall with water and fire until the vessel glows the red color of Chinese strawberry 楊梅 (*Myrica rubra*, Sieb. et Zucc.).[57] Cool it, extract the contents, and discard the beans. Grind the residue to powder, and take doses of one *ch'ien* (4 g) in warm wine.

6. Pox with sunken lesions 痘瘡陷伏 (= smallpox?). This is one of the group of diseases dominated by smallpox; its peculiarity

[55] The *tzu* (literally, "character") is a special unit of weight used in pharmacology and alchemy. It is defined in medical lexica as 1/4 *ch'ien* 錢. As the name indicates, it was originally a term for the amount of powdered medicine that would cover one of the four characters printed on the faces of common copper coins.

[56] This recipe is cited as drawn from *T*, but we do not find it there. More complex prescriptions using human and animal cranial bones are given, for example, *T 52:* 1615, which uses human bregma and the skulls of macaque and tiger as well as mercury and arsenic compounds.

We omit the next heading, "Blindness due to glaucoma 青盲不見." There is no prescription, merely a cross-reference to the article on camphor 龍腦; but no prescription using bregma is found there.

[57] In other words, the bones and beans are sealed and heated in a water-cooled cyclic reaction vessel of the sort described in Ho Ping-yü and Joseph Needham, "The Laboratory Equipment of the Early Mediaeval Chinese Alchemists," *Ambix*, 1959, *7:* 74–81. The authors note that the function of the water-cooling reservoir, usually at the top of the vessel or extending down inside, was to control the internal temperature. This is of course true, but no less important to the alchemist was the expectation that fire (yang) below and water (yin) above would interact to produce an unending cycle of condensation and vaporization, for it was by cyclic process that elixirs were perfected. This consideration is not inapplicable to the present pharmaceutical preparation, for the text emphasizes the cyclical movement that the ingredients are expected to undergo.

is the failure of the vesicles to fill with fluid to form pustules. There are several varieties, depending upon the color of the lesions (*W* 683, s.v. 痘瘡倒陷).

[Lesions] gray, flat, and stable 灰平不長. Emotions troubled and breathing rapid. Take Supernal Sentience Cover, roast, and grind. Take three *fen* (28 g) with wine. Another prescription adds two *fen* of realgar (As_2S_2); the pustules spontaneously erupt 自然起發.

7. Soft chancre 下部疳瘡. This disorder is directly attributed to venereal contact. Unlike syphilis, chancroid must have been present throughout the world from an early time. Its causative agent, the Ducrey bacillus, is from the same family (*Hemophilus*) as the agents of influenza and pertussis (whooping cough). It begins with painful urination and a pink or red sore with a shiny surface. At the next stage there is exudation of "unclean material." If chronic, the sore spreads and ulcerates, sometimes leading to the loss of part or all of the organ (*W* 16–17). The simpler of two prescriptions is given below.

Supernal Sentience Cover is scorched and ground to powder. [The ulceration] is first washed clean with an infusion of Amur cork-tree bark 黃蘗湯 (*Phellodendron amurense,* Ruprecht) and the powder is applied. Miraculously effective.

8. Wet degenerative ulcerations on the calf 臁瘡濕爛

Two *ch'ien* (7 g) of roasted and ground bone from the top of the human head 人頂骨, three *ch'ien* of dragon bone 龍骨 (various fossil bones), and one *ch'ien* of "golden thread" sulfur 金絲硫黃 (S, monoclinic) are powdered. First take Chinese radish root dug up in winter 冬蘿蔔英 [=䕅] (*Raphanus sativus,* L.) and dried in the shade, simmer in water, and wash the ulcers with the liquid.[58] Then plaster on the powder.

9. Baldness in infants 小兒白秃

This is a specific disease in Chinese medicine, with white scale that probably corresponds to that of psoriasis or dandruff. But if chronic, infected scabs eventually form. When they are bathed away large pits remain, which discharge pus. There is

[58] One suspects an element of sympathetic magic here, since a common name for Chinese radish root is "earth skull 地骷髏."

no pain and only slight itching. Ch'ao claims that almost imperceptible worms are sometimes seen in the pits. These are very likely incidental, and the more serious infections may be complications (*C 50:* 265; *Υ* 169).

Equal parts of soya beans and skull bone 髑髏骨, both roasted to ash, are mixed with lard from salt pork 臘豬脂 and applied.

COMMENTARY

It is to be expected that a high degree of superstition would attend the use of human bone as a Chinese medicinal. The skeleton persists as a long-lasting reminder of the dead body— long after the flesh has decomposed. It follows that bones possess the necessary "strength" to heal bone disorders: beatal ings, broken bones, bone trauma resulting from falls or blows, and the effects of corporal punishment, at times an occupational hazard of outspoken officials. None of the thirteen conditions treated with human bone could possibly have been favorably affected by it. It is clear that Taoist practice—a highly rational but esoteric system of ritual distilled and structured out of folk belief—dictated the specific use of the bregma. There is no reason to believe that the chemical composition of the bregma differs in any important respect from that of bone in general. The point is that bregma, unlike bone in general, was not used to treat bone disorders but to cure a wide variety of ailments whose common element is no longer obvious within the high tradition of theoretical medicine represented by Li Shih-chen.

Item B1 presents an invocation of a sort that may conceivably have dropped out of other bregma treatments as they were transposed out of the practice of priests and magus-doctors into that of literatus-physicians. This invocation provides a most valuable clue that the function of bregma therapy may have been not magical but ceremonial. In other words, the bregma prescription itself was not believed to be automatically effective, but it had to be accompanied by a petition to a higher power. The *control* of the disease agent was to be exerted by Father Thunder and Mother Lightning. The final phrase of the invocation, conventional in the extreme, reminds us that these two natural forces are constrained because the practitioner has correctly oriented himself in relation to the cosmic order to which they are also subject. Such rituals of cosmic orientation are well known. They involve facing in certain directions, pacing off the stars of certain constellations, emplacing objects with directional associations to mark out a microcosm, and so forth.[59] Whichever means was to be used in this worm treatment is no longer specified by Li Shih-chen.

It would not, in fact, be amiss to consider in subsequent investigations the possibility that the bregma was originally used *in the orientation ritual,* and that its compounding with familiar drugs and ingestion were introduced at some stage in the adaptation to the high tradition of medicine. One can see a parallel, for instance, with the use of gold made from cinnabar in early Chinese alchemy. Emperor Wu of the Han (133 B.C.) was convinced by a magician that if he ate off vessels fabricated of this gold he would attain not only somewhat greater longevity but also contact with the immortals, and that this contact and the carrying out of certain sacrifices to the forces of nature would procure the Emperor immortality.[60] This application of transmuted cinnabar seems bafflingly indirect if viewed in the light of later alchemy, where alchemical gold produced immortality by a process quite analogous to that of curing disease with medicines. To a student of anthropology familiar with ritual, however, it appears practically commonplace.

As a living tissue, bone is a collagenous protein matrix impregnated with mineral salts, especially phosphates of cal-

[59] Taoist orientation rituals are mentioned in Max Kaltenmark, *Lao tseu et le taoïsme* (Paris, 1965), p. 165 et passim; the excellent English translation by Roger Greaves (*Lao Tzu and Taoism,* Stanford, Calif., 1965) omits Kaltenmark's remarkable selection of illustrations. See also Holmes H. Welch, "The Bellagio Conference on Taoism," pp. 126 et passim. The best available ethnological study of direction concepts is Marcel Granet, *La pensée chinoise* (Paris, 1950), pp. 86–114.
[60] Holmes Welch, *The Parting of the Way. Lao Tzu and the Taoist Movement* (London, 1957), pp. 99–100. The revised paperback reprint of this work is entitled *Taoism. The Parting of the Way* (Boston, 1966).

cium. The bulk of the bone used in Chinese medicine necessarily was obtained from corpses and was almost invariably processed by roasting. Consequently, only the mineral content was important therapeutically, the protein matrix having been destroyed either by decomposition or by heat. Mineral in bone is mostly in the form of hydroxyapatite: $Ca_{10}(PO_4)_6(OH)_2$. Sodium and small amounts of magnesium, chlorine, copper, carbonate (as CO_3), fluorine, iron, manganese, and strontium are also present.[61] The administration of bone in cases of bone fractures or trauma (items A2 and A4) cannot affect the course of bone healing or the recovery from physical trauma. In the case of fractures, successful alignment of the fragments is all-important for proper healing; the calcium salts for mineralization of the healing bone are derived either from the injured bone itself, the body fluids, or the diet, which normally has an overabundance of calcium salts and phosphates. The ingestion of human bone would be superfluous except in rare cases of calcium deficiency, and would not affect the healing process.

None of the first six entities for which bregma was used can be specifically identified with any degree of confidence. To cite a few examples: item 5, inability to eat, might be anything from anorexia nervosa to esophageal stricture; item 6, "pox with sunken lesions," could be any of the infectious exanthematous diseases or a host of other disorders; item 4, "autumnal intermittent fevers," might include malaria, typhoid fever, brucellosis, or even influenza. Whether "cadaver vector disease" really meant pulmonary tuberculosis, as seems likely, or some other debilitating infectious disease is really not pertinent to this discussion, because bone would be ineffective in the treatment of any infectious process. By the same token, bone would not alter the course of soft chancre or of any other type of genital infection. Baldness in infants, usually caused by impetigo—skin infection with *Staphylococci* or *Streptococci*—cannot be cured by the application of bone prepared with soya bean and lard.

[61] See Harrison et al., eds., *Principles of Internal Medicine*, pp. 690–696.

Wet ulcers on the calf were treated with bone and elemental sulfur. This is an example of an additive's having a greater curative effect than the human drug itself. Sulfur is still widely used as a desiccant and antieczema agent by dermatologists.[62] The topical application of sulfur to weeping leg ulcers would tend to dry them and probably retard bacterial growth, thereby hastening the normal healing process.

Conclusions

The basic question prompted by our study is not so much medical as historical and philosophical: what were the crucial factors that determined the incorporation and retention of drugs in the Chinese pharmacopoeia? Were these factors pharmacological, theoretical, magical, or some combination of the three? Although this study is meant more to encourage methodologically sound exploration on many fronts than to provide definitive answers, two conclusions are supported by the evidence at hand. First we shall state them as negative generalizations, review the supporting data drawn from our study of human drugs, and then examine factors that will affect the validity of our conclusions over the whole of traditional Chinese drug therapy.

1. No single factor among the three—known pharmacodynamic effects, symbolic procedures established within folk medicine, the theoretical abstractions of the ancient scientific traditions—can account for the range of prescriptions at the old-fashioned physician's disposal. The positivistic conception of Chinese drug therapy as an empirical science analogous to the clinically tested therapeutic procedures of modern medicine does not stand up to close and impartial scrutiny. Unfortunately, it is too useful for the purposes of nationalistic propaganda to be lightly discarded or heavily qualified. Anyone whose concern is to comprehend traditional medicine as an entity subsisting and changing through time will, we believe, find it a waste of time to confine his attention to what looks interesting

[62] See A. L. Welsh, *The Dermatologist's Handbook* (Springfield, Illinois, 1957), p. 44.

from the positive point of view. The crucial problem is rather to understand the balance between empirical, ritual, and theoretical factors, and to discover the many modes of their interplay.

2. The role of clinical observation in the continued use of drugs or combinations of drugs over millennia by the Chinese practitioners was insufficiently developed to yield a sound general understanding about either the specificity of individual drugs or the active ingredients of prescriptions. While we do not wish to deny that differential trials with the various ingredients of prescriptions led in certain cases to a comprehension of their functions, we have seen that in the cases discussed above efficacy depends as often on an ancillary ingredient or a synergistic effect on the ingredient under which the prescription is classified in the pharmacopoeia.

More fundamental, pharmacologic factors could not be systematically differentiated from ritual factors in drug classification. The highly schematic character of Chinese medicine required that every concept, every measure, every agent, be related indirectly or directly in the nexus of theory. The criteria by which drugs were fitted into the medical literature and their action explained were extremely abstract.[63] Drugs that had entered medical practice for reasons that we would not consider specific—for instance those whose action depended upon sympathetic magic—could be theoretically rationalized just as easily as highly specific drugs, and thereafter would be no less difficult to dislodge so long as a certain number of patients recovered after taking them.

Tables 8.1 and 8.2 review the evidence for these conclusions. Fewer than five percent of the disease entities considered could have derived specific benefit by known effects of the main ingredients of prescriptions, and only about three percent could have been helped by known effects of the ancillary ingredients. Thus less than eight percent of all these disorders

[63] There is no satisfactory account of the conceptual and classificatory system of Chinese materia medica in a Western language, but some of its outlines may be glimpsed in Lu Gwei-djen, "China's Greatest Naturalist," and in Bretschneider (see note 8, this chapter).

Table 8.1 Analysis of the Possible Medicinal Value of the Ingredients of the Chinese Human Drug Prescriptions

	Number	Percentage
Total number of diseases listed in Pharmacopoeia*	81	—
Total number of diseases considered†	66	100
Total number of diseases *not* benefited by the human drug in question or by any ancillary ingredients	50	75.8
Total number of diseases possibly benefited by the human drug *or* one of the ancillary ingredients	16	24.2
Possible benefit from human drug alone	8	12.1
positive benefit	3	4.5
uncertain or nonspecific benefit	5	7.6
Possible benefit from ancillary ingredient alone	5	7.6
positive benefit	2	3.1
uncertain or nonspecific benefit	3	4.5
Benefit from use of combination of human drug together with ancillary ingredient	3	4.5

Note: Of the 66 entities considered, 50 (75.8 percent) were treated by combining the human drug with one or more ancillary ingredients. In five such cases (only 10 percent of the total number of diseases treated with combination therapy), the coingredient was probably helpful in curing the disease or in alleviating the symptoms.
*Treatable by the eight human drugs (with or without admixtures) under consideration.
†Fifteen diseases were omitted from the discussion because they could not be identified or because they were not related in the text to medical therapeutics.
We do not claim wider validity for these statistics, since the human drugs are not representative of the pharmacopoeia as a whole.

could have been positively cured by known pharmacological effects of the remedies cited. If one includes nonspecific therapy, the highest percentage of known benefit is twenty-four. This figure includes possible therapeutic effectiveness resulting from synergism between human drugs and coingredients. The cases of known positive and specific benefit (items V.B 1–4) are concentrated in one category, that of nonmenstrual human blood.[64] Even so, the use of blood was in all probability origi-

[64] In modern medicine these would more likely be considered symptoms than diseases. At least two of the four, nosebleed and postpartum nausea, would probably remit spontaneously sooner or later.

Table 8.2. Entities Where Ancillary Ingredients Were Used in Prescriptions

		Total Number
I	4, 5, (6)	3
II	3, 4, 5, 8, 9, 10, (11), 12, 13, 14, 15, 16, 17	13
III	1, 2, 3, (4)	4
IV	*1*, 2, 3, 4, 5, *7*, 8, 9, 10, (11)	10
V	A=2, 3, 4, 5, 6, 7, 8, 9, 11 B=1, *4*, 6	12
VI	1, 4	2
VII	1, 2, 4	3
VIII	A=1, 2, 4 B=1, 3, 4, 5, 6, 7, (8), 9	11
		58

Notes: The Roman numerals followed by Arabic numbers correspond to the prescription numbers in the text.

It is clear that in the great majority of prescriptions, ancillary ingredients were combined with the human drug. As indicated above (Arabic numerals enclosed in parentheses), only five of the 58 entities so treated could have benefited from the coingredient *alone*. Three instances where benefit was likely through the combination of the coingredient with the human drug are italicized. Thus, eight (16 percent) of the 58 entities considered here could possibly have been aided by the coingredient itself or by the coingredient in synergism with the human drug.

nally dictated by what we would now call replacement therapy, a rationalization of sympathetic magic. The use of blood to treat blood loss does not differ in principle from the positivistically inexplicable attempt to treat bone damage with bone (items VIII. A 2 and 4).

Instances where drugs or coingredients were *possibly* beneficial from a scientific viewpoint are rare.[65] Nonspecific assistance from human drugs is possible in the following cases: human nails in treatment of open wounds (II.3), foreign bodies in the eye (II.13, 14), epistaxis (II.18), and human milk for insects in the ear (IV.14). Ancilliary ingredients are of clearly predictable benefit in only two instances (see Table 8.2): tung oil for calf sores (IV.11), and sulfur for weeping calf ulcers (VIII. B 8). They are of possible benefit in the following instances: soap-bean thorns for felons (I.6), silkworm cocoons for ulcerated

[65] Of course we do not consider modern pharmacology a complete and finished body of knowledge. We expect that in the future more discriminating judgments about specific properties of human drugs will be possible; nothing in this study should be misinterpreted to exclude that possibility.

hemorrhoids (II.11), and calomel for fistulary lymphadenitis (III.4). Combinations of human drugs and ancillary ingredients are of known possible benefit in the following cases: human milk and human placenta for debilitation due to *ch'i* exhaustion (= anemia?; IV.1), human milk and copper for trachoma or other eye infections (IV.7), and human blood plus vinegar for treatment of postpartum nausea due to blood loss (V.B 4; see Table 8.2).

It is evident from Table 8.1 that, if the ingredients of the Chinese prescriptions had been chosen by the most stringent criteria of clinical pharmacology, fewer than one-quarter of the ailments treated could have been relieved strictly by known properties of the constituents or their combinations. In other words, if Chinese drug therapy had had the scientific basis sometimes claimed for it, and if cures depended upon scientific basis alone, failure of seventy-five percent of the prescriptions would have resulted in their elimination as time went on. We must emphasize the necessity for a serious address to the complex problem of why this elimination did not take place. Even in the most scientifically advanced countries the process of discarding unverifiable remedies is anything but automatic.

We have suggested that the repertory of drug therapy evolved by a process that reflects the social character of traditional medicine in China. On the one hand, there was the folk medicine of the illiterate villager and town dweller, practiced by local or itinerant healers. Some of these knew a few drugs well and used them intensively. Some, priests, shamans, spirit mediums, and peripatetic wizards among them, used ritual and magical therapies. Their patients believed in them, for the uneducated and slightly educated majority throughout history have lived in a world full of spiritual agencies, governed by capricious cosmic forces whose sufferance could be ensured only by endless symbolic acts and propitiation.

The medicine of the "great tradition," on the other hand, resembled the other sciences of the elite minority. Although the vitality and potentiality for growth of high pharmacology depended heavily on the discoveries of illiterate practitioners, it imposed upon them a remarkably abstract and unitary structure of theory. The experience of the village was truly transposed into another world, where what one had to fear was not hungry ghosts and pestilential demons, but defects in the utterly impersonal rhythmic balance of yin and yang and the Five Phases. The documents of this high tradition were compiled with the goal of earning a place in a long succession of canonic texts. Although many eminent scholar-physicians deserve credit for substantial clinical contributions of their own, most of the content of the therapeutic classics was systematic compilation from earlier literature and hearsay[66] and thus contrasts with the intensely pragmatic knowledge of the folk practitioners. One might say that the scholar-doctor usually knew much more than he needed, and the folk healer seldom knew enough.

The division between the two medicines was never absolute. Their epistemology differed fundamentally, but they shared, as any medical science must, a responsibility to the patient and thus to clinical efficacy in its widest sense (without separate conceptions of psychogenic and psychosomatic disorders, there was no reason to value a ritual or placebo that cured the patient more or less than a specific drug). Despite the great abstracting thrust of classical medicine, its compilers needed the richness

---

[66] Recent attempts to portray Li Shih-chen as a great empiricist have tended to oversimplify the picture by ignoring frequent evidence of his credulity, as in this example of a human drug (L 52: 109, s. v. "penis," jen shih 人勢). "The human yin stalk is not a medicine. But T'ao Tsung-i 陶宗儀 in his Cho keng lu 輟耕錄 (1366) relates that in Hangchow there was a Mr. Shen who had committed rape and was found out. He took a knife and castrated himself. He bled for months, and the wound did not heal. Someone had him find the penis which he had cut off, pound it to powder, and take it with wine. In not many days he had recovered. Contemplating this story, it would seem that those 'who go down to the silkworm room' [who are administratively sentenced to castration] should not be ignorant of this method, so I append it here."

We have not dealt with this human drug in this chapter because no prescriptions are appended. The anecdote cited by Li can be found in the Chin tai pi shu 津逮秘書 ed. of Cho keng lu, 9: 7b.

of folk medicine too badly to filter out its peculiar symbolic content entirely. The fertility of many of the great medical authors came, we suspect, from their ability to straddle the two worlds and take them both seriously. Symbolic procedures, once transposed from their milieu into the abstract realm of rational medicine, posed no threat. Doctors who maintained the strictest Confucian skeptical standards could ignore anything in the medical literature that they considered superstitious. Literate physicians whose milieu lay close to the borderline between these two realms could make extensive use of amulets and incantations, comprehending them in rather abstract terms.

That there were doctors on the ideological borderline implies a continuum, rather than a dichotomy, between the elite and folk medical traditions, even though the styles and values of the extremes remained distinct in important ways. We know from written records that physicians were found at practically every point along the spectrum of social standing and education, even though the classics were necessarily written by men of some position (Li Shih-chen's grandfather was an itinerant doctor, but his father and he, having passed the local civil service examination, were formally members of the gentry).[67] Recent field studies have made it obvious to anyone who has studied the high tradition that a great deal of rational medicine diffuses downward into the conceptions and beliefs of uneducated people.[68] Just as in the "upward" passage the emotive and

[67] Lu and Needham, "Medicine in Chinese Culture," pp. 265–266.

[68] The most important of these available to us are Topley, "Chinese Traditional Ideas," and Francis L. K. Hsu, *Religion, Science and Human Crises. A Study of China in Transition and Its Implications for the West* (International Library of Sociology and Social Reconstruction; London, 1952). There is also much valuable information on beliefs about mediumistic healing in Alan J. A. Elliott, *Chinese Spirit-medium Cults in Singapore* (The London School of Economics and Social Science, Monographs on Social Anthropology, 14; London, 1955).
The influence of Buddhism on popular ideas about the body, health, and sickness is another area that will richly repay exploration. An excellent beginning has been made by P. Demiéville in his article "Byō" in *Hōbōgirin. Dictionnaire encyclopédique du Bouddhisme d'après les sources chinoises et japonaises* (Paris, 1937), pp. 224–265.

ritual content of disease experience is replaced by abstract theory, in the "downward" passage elements of rational medicine reenter folk consciousness with their experiential content restored and reconstructed. This circular flow implies that certain conceptions of disease an anthropologist might elicit from an informant have survived in Hong Kong or Szechuan until today (with remarkable uniformity over the whole of China, it appears) partly because indirect contact with the written tradition at the other end of the spectrum was never lost. Understanding the conceptual transformations which take place in the upward and downward passages is one of the chief keys to the history of Chinese medicine.

Now what is our evidence for this extended and very tentative hypothesis about social factors which affected the content of the classical tradition of medicine?

First, there is the very substantial magical and ritual content of the pharmacopoeia. Among the human drug prescriptions we have examined, we find obvious reasons to suggest symbolic content for a number of procedures:

I.4
II.1, 4, 10
V.A 1, 5, 7, 8, 9, 10, 11
V.B 1, 2, 3, 4, 5
VIII.A 2, 4
VIII.B 2, 4

We believe that others (particularly in groups VI and VIII.B) will be identified by those qualified to make a more sophisticated examination.

Measuring the clinical efficacy of ritual procedures must be left to future investigation. We see no conflict with modern medicine, however, in the idea that, because social factors influence a culture's definitions of disease, sickness and recovery are to some extent learned behavior. We have noted in our commentary to item V.B 4 that nausea and abnormal bleeding are quite likely to result when a woman in labor realizes that she has inadvertently violated a directional taboo; her normal dismay and its psychosomatic consequences would be reinforced by the expectations of those about her familiar with the disorder "postpartum nausea with abnormal blood loss."

Second, we have found traces of the process by which the magical content of folk medicine is denatured by assimilating it to the world view of the "great tradition." In item V.B 4, for instance, the directional taboos that women are expected to observe during labor are cast in terms of the Five-Phases theory.[69] In the introduction to section VIII we cited a passage in which Li Shih-chen found a perfectly naturalistic solution in terms of the protoscientific notion of resonance to account for a most exceptionable story about an ache cured by caring for a neglected bone chip that had been removed from the patient's leg long before. There is also the remnant of a cosmic ritual in item VIII.B. 1 to ponder.

We will add an even more paradigmatic example. In Ch'ao Yuan-fang's great nosological system (*Chu ping yuan hou lun*) there are many disease classes that reflect with particular directness the symbolic world of folk medicine. The *ku* poisoning diseases (*ku tu* 蠱毒), which have been intensively studied, are due to what Westerners would call witchcraft.[70] There are two other classes in particular, many of whose members are caused by spirit and demon possession: the "cadaver vector diseases" already encountered in this paper (pp. 251 and 257) and the "fixation diseases" (*chu* 注).[71] Ch'ao agrees that the *chu* include dangerous states of demonic possession, but his highly

[69] We are not inclined to read overwhelming significance into this particular example, for it must be remembered that some of the concepts of the "great tradition" attained great currency in everyday thought. The juxtaposition of directional taboos and Five Phases may well have been accomplished in folk medicine. What concerns us is Li Shih-chen's acceptance of this juxtaposition. On the Five Phases and resonance theory see Needham, *Science and Civilisation in China*, II, 253–265 and 280 ff.

[70] Feng Han-yi and J. K. Shryock, "The Black Magic in China Known as Ku," *Journal of the American Oriental Society*, 1935, *55:* 1–30.

[71] On spirit possession as an entity in modern psychiatry, see P. M. Yap, "Classification of the Culture-bound Reactive Syndromes," *Far East Medical Journal*, 1969, *7:* 223, and "The Possession Syndrome: a Comparison of Hong Kong and French Findings," *Journal of Mental Science*, 1960, *106:*114 ff.

conceptual explication is concerned only with the details of the somatic dysfunction that makes it possible for the possession to be established. Nothing remains in the passage from which this excerpt is drawn to account for the terror felt by the victims of "fixation diseases" (*C 24*: 130a):

When we say that *chu* has the sense of "stay" (*chu* 住) we mean that a pathological *ch'i* takes up residence within the human body; thus it is called *chu* 注.[72] This comes about because the [somatic] yin and yang do not discharge their responsibilities as they should, the *ching* 經 and *lo* 絡 circulation channels become empty, and the disease is brought on by internal wind, cold, heat, damp, overexertion and fatigue factors. . . .

Preeminent among scholar-physicians who could accept symbolic practices without obvious rationalization was Sun Ssu-mo, active through most of the seventh century. We find incorporated in his *Ch'ien chin fang* and *Ch'ien chin i fang* 千金翼方 a trove of amulets, incantations (even in Sanskrit), and all sorts of other practices that obviously derive from folk religion. As a Taoist and an alchemist Sun was better prepared than Ch'ao to accept on equal terms the worlds of folk belief and high theory. Sun's cure for possession by ghosts is two series of exorcistic formulae that are quite unambiguously meant to be performed by a Taoist priest.[73] His successor Wang T'ao, who consistently depends upon Ch'ao Yuan-fang for etiology, produces a number of matter-of-fact-looking prescriptions for

---

[72] Ch'ao's somewhat forced use of a homonym is conventional in Chinese medicine, and in lexicography generally. He explains later that ghosts and demons are numbered among these "pathological *ch'i*."

[73] *Ch'ien chin i fang* (reprint of 1878 ed.; Peking, 1955), *29*: 347 and *30*: 354–355. Both formulae in the first series correspond in detail to Taoist rituals. In the first the exorcist speaks of himself as "wearing the seal of authority at my waist and the Glory up on my head, and with my feet pacing out the ambit of the constellation K'uei"; in the second he begins "I am a Libationer of the Celestial Master [Sect], emissary of Sky and Earth." On pacing out stellar ambits *("pas de Yü")*, see Kaltenmark, *Lao tseu et le taoisme*, p. 165. As is well known, the Celestial Master sect began as a healing cult. The second series, for "fixation diseases" (*tun chu* 遁注), includes ten formulas that make much use of name magic.

"fixation disease." But most of them incorporate extremely poisonous substances from all three realms of nature. One, for instance, combines cinnabar, arsenolite, realgar, croton seed (*Croton tiglium*, L.), black veratrum root (*Veratrum nigrum*, L.), aconite (*Aconitum*, L.), and centipedes. In addition to ingesting the compounded pills, the patient is told to carry one pill on his person to keep away evil! It is very much to the point that a similar formula found in earlier alchemical literature is not administered to a patient but burnt as a demon-killing incense.[74] Li Shih-chen himself does not often impose rationalization on magical material he accepts from his sources, but occasional comments such as those we have cited (Sec. VIII, Introduction) indicate his implicit tendency toward abstraction.

In assessing the limitations of this investigation, we must return to the fact that the drugs considered in this study were chosen because of a single common characteristic: they are of human origin. They were not selected to serve as a representative sample of the Chinese apothecary's stock. Indeed, because of their human derivation, more superstition and magic are associated with them than with most drugs of vegetable, mineral, or lower animal origin. For just that reason, their study throws into bold relief complexities that would emerge from other areas of the pharmacopoeia only after much more extended and conceptually more rigorous explorations. A preliminary survey indicates that obvious symbolic procedures are far from rare elsewhere, although as in the case of the human drugs they tend to cluster about certain substances more than the rest. Although our conclusions are, we believe, firmly established on the basis of the evidence presented, the general validity of our speculative hypotheses can be affirmed or denied

[74] *W 13:* 362b. All of these ingredients except arsenolite also appear in a demon-killing *incense* formula given in an alchemical tractate attributed to Sun Ssu-mo, the *Tan-ching yao chueh* 丹經要訣 (Essential formulas for oral transmission from the alchemical classics); see Sivin, *Chinese Alchemy,* pp. 208–209, and the discussions of individual components in Appendix G.

only by further systematic study.[75] We would suggest, as a second step, studying a wide variety of prescriptions and treatments for a single group of diseases, in order economically to identify other substances rich in symbolic associations. The greatest yield in additional knowledge would perhaps come from a group in which psychogenic factors are prominent—for instance, disorders involving possession, fright, or sexual disability. Another fruitful area for exploration would be that of epidemic illnesses. In the small tradition, because epidemics affect a considerable portion of the community at once, they especially tend to be treated by the communal rituals of folk religion.[76]

We will close with a few remarks on the built-in bias that results from the use of one medical system to evaluate another. Modern medicine may be considered culture-free to the limited extent that its methods of verification are public, rigorous, and capable of yielding identical results when applied to patients in any culture or social milieu regardless of their faith or even confidence in the therapy being tested. At the same time it has a cultural bias toward values and attitudes that have been characteristic of the West since the Scientific Revolution, and that are very gradually affecting the rest of the world. This bias becomes particularly visible in areas that, while having proved their pragmatic value in the broad attack on human disease, are only gradually being subordinated to objective and rigorous criteria of clinical success and physiological comprehensibility: in particular, psychiatry and psychosomatic medicine. We hardly need emphasize again the importance of

[75] Our generalizations may be said to have some validity for the period from which Li Shih-chen draws his detailed prescriptions (mainly from the T'ang on). The relation between magico-religious and empirico-rational medicine in the Ch'in and Han has been studied by Miyashita Saburō 宮下三郎, "Chūgoku kodai no shippeikan to ryōhō 中國古代の疾病觀と療法" (The ancient Chinese view of disease and therapy), *Tōhō gakuhō* 東方學報(Kyoto), 1959, *30:* 227–252.
[76] Hsu, *Religion, Science, and Human Crises*, esp. ch. 5. One can gather from Hsu that what characterizes religious healing in China, regardless of the disorder brought to it for cure, is communal expiation.

these two branches to the comprehension of Chinese medicine, and the danger of applying their insights destructively.

The fundamental concepts from which the two systems—Chinese and modern—arise are unrelated and irreconcilable.[77] The only fundamental *theoretical* similarity between the two systems is that the therapeutic approach to disease takes into account the body as a whole; it makes no sense to treat an isolated part and ignore the rest of the body. Modern medicine evolved in the closest possible consonance with a scientific method that originated in physics. Chinese medicine is a rational construction originating from basic conceptions of the universe and its microcosm, man. Data taken from experience were systematically worked into a metaphysical structure that could be neither buttressed nor destroyed by experimental proof. Disease was thought to be caused by fundamental disruption, due to either external or internal causes, of the flow of energetic *ch'i* and vital nutriment, which in the healthy body would be balanced, equitable, and in phase with the cosmic rhythms of the day and year. In diagnosis, identification of a disease "vector" was usually less important than tracing the locus and progress of the pathological *ch'i* imbalance. In the West, probably mainly since the discovery of microbes, we tend to discount the existence of disease agents other than those whose existence can be rigorously demonstrated. Western medicine would rather advocate therapeutic nihilism in cases where the disease is of unknown origin than commit itself to occult causation. Chinese doctors were not concerned with the dichotomy of scientific and supernatural healing powers. Many would be as scornful of "superstitious" healing rituals as a new graduate of an American medical school. But few would deny outright the efficacy of those rituals, and most would accept the incorporation of therapeutic measures borrowed from them into the classical literature.[78] Once these measures

[77] See the essay review by Sivin in *Journal of the American Oriental Society*, 1968, *88:* 641–644.
[78] Lu and Needham ("Medicine and Chinese Culture," p. 268) note that under the Sui (A.D. 585) the Directorate of Medical Administration includ-

were tied, loosely or tightly, into the rational structure, the
stigma of superstition or heterodoxy no longer applied (there
is often an element of snobbery or social prejudice in Chinese
judgments about heterodoxy). Attribution of a cure to restoration of the normal *ch'i* circulation would be identical regardless
of the therapy's origin or character. If a Western doctor happened to cure the same patient's disorder by nonspecific means
(and he would consider acupuncture and moxibustion no more
specific than amulets and incantations), he would most likely
avoid speculation or attribute the cure to suggestion, bedside
manner, or some other kind of rough-and-ready psychotherapy.
Thus both the traditional and the modern doctor would interpret the cure according to the training and medical heritage
of his society. Neither doctor's interpretation can at the moment
be proven; the metabolic processes by which psychotherapy
affects a cure are as obscure as those proposed for acupuncture
and moxibustion.

It is obviously of momentous concern that the study of the
mental and emotional components of disease be freed from
cultural limitations. This is bound in principle to happen
sooner or later, whether because every aspect of medicine is
successfully brought under the dominion of modern experimental science or because the homogenizing drive of mass consumer
society obliterates fundamental local distinctions of behavior,
attitude, and value. But at the present moment the con-
summation of One World is less likely than either global

---

ed two professors of apotropaics (that is, ritual medicine). They do not
mention that ritual medicine was imperially sponsored in this manner
throughout the ensuing thousand years until the end of the Ming, according
to the documents conveniently collected in Ch'en Pang-hsien 陳邦賢,
Chung-kuo i-hsueh shih 中國醫學史 (A history of Chinese medicine; reprint,
Taipei, 1958), pp. 125–142 and 200. Lu and Needham's assertion that the
use of charms, incantations, and invocations was "quite peripheral to the
practice of medicine as such, kept far indeed from the center of the stage,
and it can confidently be asserted that from the beginning Chinese medicine
was rational through and through" is contradicted by an overwhelming
body of evidence, of which we have offered only a tiny selection in this
paper. The fundamental rationality of Chinese medicine lay in its theory,
not in its practice.

incineration or incessant convulsion, and a new scientific revolution that strictly correlates the behavioral manifestations and mental content of psychopathology with its biochemistry is nowhere in sight. In the meantime comparative study offers tools by which the universal characteristics of medicine may be distinguished from what is merely local or historical accident. The richness, sophistication, and accessibility of the traditional Chinese healing art make it an unsurpassable starting point for building a comparative history of medicine.

9

A Neglected Source for the Early History of Anesthesia in China and Japan

Saburō Miyasita

From early times there have been crude attempts to diminish the pain of surgical operations. In China the *ma-fei san* 麻沸散 of Hua T'o 華陀 (fl. first half of the third century A.D.) is thought to have been Indian hemp.[1] Mandrake once traveled to Yuan China as an anesthetic,[2] and the first operation under sulfuric ether was performed by Peter Parker, Yale graduate and medical missionary, in 1847 in Canton.[3] In Japan, the famous surgeon Hanaoka Seishū 華岡青洲 performed an operation using an anesthetic made from white datura in 1805,[4] and modern anesthesia was first introduced by the Dutch physician Pompe van Meerdervoort, who arrived in Nagasaki in 1857.

In this paper we describe a method of anesthesia developed and established in the Yuan dynasty (1279–1367). Chinese surgeons appear to have used white datura, among other drugs, more than four and a half centuries before Hanaoka, in

[1] Wong Chi-min and Wu Lien-teh, *History of Chinese Medicine, Being a Chronicle of Medical Happenings in China from Ancient Times to the Present Period* (second ed., Shanghai, 1936), p. 55. The identification of *ma-fei san* is substantiated in Kung Ch'un's "A Great Chinese Surgeon, Hua T'o" [in Chinese], *Chung-hua i-shih tsa-chih* 中華醫史雜誌, 1955, *7:* 24–28. See also Chapter 8, p. 248, in this book.
[2] Kumagusu Minakata, "The Mandrake," *Nature,* 1896, *54:* 343–344; Saburō Miyasita, "Mandrake Once Traveled to China as an Anaesthetic," *Japanese Studies in the History of Science,* 1966, no. 5, pp. 189–192.
[3] Wong and Wu, *History of Chinese Medicine,* p. 339.
[4] Kure Shūzō, *Hanaoka Seishū sensei oyobi sono geka* 華岡青洲先生及びその外科 (Dr. Hanaoka Seishū and his surgery; Tokyo, 1923); Hiromu Takebayashi, "Seishu Hanaoka, Pioneer of the General Anesthesia and the Modern Surgery," *Japanese Studies in the History of Science,* 1967, no. 6, pp. 115–123; W. N. Whitney, "Notes on the History of Medical Progress in Japan," *Transactions of the Asiatic Society of Japan,* 1885, *12:* 313; Yū Fujikawa, *Japanese Medicine* (tr. from the German by J. Ruhräh; Clio Medica Series, no. 12; New York, 1934), p. 57.

what may be the earliest documented surgical procedure under general narcosis in the world.

Man-t'o-lo hua 曼陀羅花 (literally, "mandala blossom") is the dried fruit and flowers of white datura (*Datura alba*, Nees), native to tropical Asia. It contains scopolamine, an important anesthetic in modern medicine, and smaller amounts of hyoscyamine and atropine.

It is well known that Li Shih-chen 李時珍, in compiling the Great Pharmacopoeia (*Pen-ts'ao kang mu* 本草綱目), which was published in 1596, reported that he had experimented with white datura in order to investigate its effects on behavior as well as on pain. In ch. 17 he wrote:

There is a tradition that, if you laugh as you gather these flowers and use them in brewing wine, it will make people laugh; if you dance as you gather them, it will make people dance. I have tried it. The subject should be given enough of the wine to make him half-drunk. If then you have someone else laugh or dance to lead him on, it will work. You gather these flowers in the eighth month, and add equal parts of the flowers of Indian hemp gathered in the seventh month. Dry them in the shade and then powder them. Shortly after taking three *ch'ien* 錢 (2.5 g) of the powder with heated wine, you will be confused, as though drunk. The powder may be taken prior to cutting out abscesses and burning moxa, so that you will not feel the pain.

In the Chinese cultural sphere, the physiological properties of white datura were first noted in 1045 when, while quelling an insurrection in Kwangsi, the commanding officer Tu Ch'i 杜杞 used wine in which white datura had been steeped as a narcotic.[5] The idea probably came from local folk knowledge. In 1178, in the *Ling-wai tai ta* 嶺外代答, Chou Ch'ü-fei 周去非, an official of Kweilin (Kwangsi), reported that a certain wine of Chao-chou 昭州 was very intoxicating because it contained the spirit of white datura.[6] He also noted that thieves used the powdered herb as a narcotic.[7] In 1082 Chou Hsu 周叙 described three varieties of the flowers cultivated in the gardens of

[5] *Sung shih* 宋史, "Biography of Tu Ch'i," *300:* 11b.
[6] Section on wine, *6:* 7a.
[7] Section on white datura, *8:* 7a.

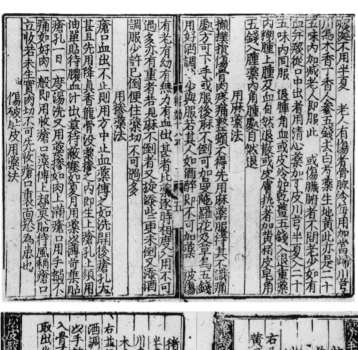

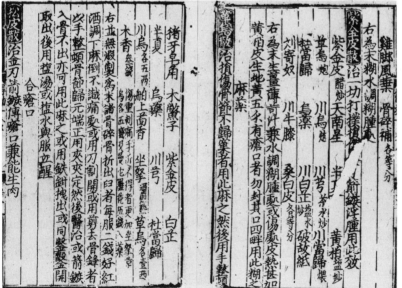

Fig. 9.1. From *chüan* 18 of *Shih-i te-hsiao fang,* second edition. Courtesy Ishizaki Collection, Osaka Prefectural Library. Section A (above) is entitled "Method of Administration of an Anesthetic"; Section B (below) is headed "The Anesthetic."

Loyang.[8] This suggests that white datura was still rare in the northern part of China. Although natives of South China were aware of the narcotic properties of the plant, its surgical applications were never discussed in the Sung period.

We first find a surgical anesthetic containing white datura in the *Shih-i te-hsiao fang* 世醫得效方 (Effective prescriptions for hereditary physicians), written by Wei I-lin 危亦林, professor in the medical school of Nan-feng 南豊 (Chekiang) and published in 1343.[9] In *chüan* 18, "Bonesetting and Treatment of Wounds," we read:

Method of Administration of an Anesthetic. When the patient is in such pain from a fracture due to a fall that it cannot be set, give him one dose of the anesthetic (*ma-yao* 麻藥).[10] Wait until he can no longer feel pain in the affected part before performing the surgical operation. If the patient still does not lose consciousness, let him take slowly a cup of good wine with five *ch'ien* (6.25 g) each of white datura and aconite. No more should be administered after he appears intoxicated.

The quantity of white datura and aconite must be adjusted to age, state of debilitation, and loss of blood, and excess avoided. If the patient does not lose consciousness after an increased dose, give him a bit more wine with white datura and aconite, but avoid an excess.

The Anesthetic. A powder containing aconite (*ts'ao-wu san* 草烏散); cures injuries to bones and joints, and dislocations. Give this powder to make the patient insensible of pain and then set the bone by hand.

Fruit of *Gleditschia officinalis*, Hemsl. (*chu-ya tsao-chiao* 猪牙皁角);[11]

---

[8] *Lo-yang hua mu chi* 洛陽花木記 (in *Yuan-pen shuo fu* 原本説郛, vol. 26), p. 18b, s. v. "Flowering Plants."

[9] This work was completed in 1337. The first woodblock edition was published in 1343 in Chien-ning, Fukien, and the second, which is very similar to the first, in 1506. A Korean edition was published in 1425, and one in movable type about 1484. There was no Japanese edition.

[10] This is one of the earliest uses of the term *ma-yao* (*mayaku* in Japanese) for "anesthetic." Today it is the ordinary medical term in China and Japan.

[11] These identifications are based on *Chung-yao chih* 中薬志 (4 vols., Peking, 1959–1961) except *Scopolia sinensis* for *tso-na*, which is my own hypothesis.

Seeds of *Momordica cochinchinensis,* Spreng. (*mu-pieh-tzu* 木鱉 [or 鱉] 子);
An unknown herb (*tzu-chin p'i* 紫金皮);
Root of *Angelica dahurica,* Benth. et Hook. (*pai-chih* 白芷);
Rhizome of *Pinellia ternata,* Breit. (*pan-hsia* 半夏);
Root of *Lindera strychnifolia,* Vill. (*wu-yao* 烏藥);
Rhizome of *Ligusticum wallachii,* Franch. (*ch'uan-hsiung* 川芎);
Root of *Angelica decursiva,* Franch. (*t'u tang-kuei* 土 [誤作杜] 當歸);
Root of *Aconitum carmichaeli,* Debx. (*ch'uan-wu* 川烏):

5 *liang* (62.5 g) each.

Imported fruit of *Illicum verum,* Hook fil. (*po-shang hui-hsiang* 舶上茴香);
Root of *Scopolia sinensis,* Hemsl. (*tso-na* 坐拏), extracted with boiling wine;
Mature aconite root (*shou ts'ao-wu* 熟艸烏):

1 *liang* (12.5 g) each.

Root of *Saussurea lappa,* Clarke (*mu-hsiang* 木香):

3 *ch'ien* (3¾ g).

In cases of serious injury with needlelike pains, when the patient cannot endure being touched, add five *ch'ien* (6¼ g) each of scopolia root, aconite root, and white datura to the powder.

Make a powder of all the above drugs without roasting. A patient with a fracture, shattered bone, or dislocation takes a dose of two *ch'ien* (2.5 g) of the powder with good red wine.

When the patient no longer feels pain in the affected area, if necessary the affected part is cut open with a knife and any sharp bone ends trimmed away with shears. Then set by hand in the correct original position and bind to a splint. In case an arrowhead has pierced the bone and cannot be pulled out, the patient is rendered insensible by this anesthetic. After the arrowhead is pulled out with iron nippers or wedged out with a chisel, the patient is immediately brought to his senses by giving him hot or cold salt water.

This was a key achievement in the history of anesthesia, although, because of the rarity of the *Shih-i te-hsiao fang,* it has not previously come to the notice of medical historians. Chinese anesthesia failed to advance further until in 1805 the

Japanese surgeon Hanaoka Seishū, who knew Dutch surgical techniques, performed successful operations under *tsūsen san* 通仙散 (Communion-with-Immortals Powder), composed of five ingredients including white datura.[12] Indeed, all five ingredients of *tsūsen san* are found among the fourteen herbs in Wei I-lin's anesthetic. A worthwhile theme for further investigation is the developments that filled the gap between Wei I-lin's record of 1343 and Hanaoka's painless operation in 1805.

[12] Hanaoka's anesthetic employed *Datura alba* (*mandarage*), *Aconitum* spp. (*sōuzu*), *Angelica dahurica* (*byakushi*), *Angelica decursiva* (*tōki*), and *Ligusticum wallachii* (*senkyū*).

An Introductory Bibliography of Traditional Chinese Science: Books and Articles in Western Languages

Nathan Sivin

Contents

Introductory Note

This bibliography is a first attempt to provide a concise but informative introductory guide for the person who wants to investigate the Chinese scientific tradition without knowing the language. Its purpose is not to point out what is worth reading on a given theme, but to guide the beginner in locating the significant ideas and developments and searching out what is known about them. Several years' use of an earlier draft of this bibliography in teaching indicates that one of the greatest needs is for guidance with respect to very general reference works and books on history and philosophy that can serve as an introduction to the background of scientific change. For this reason much space is devoted to standard documentary collections, general surveys, and bibliographies that pay due attention to science. These are useful at the stage of orientation; beyond it no reference tool can be substituted for close study of and extended reflection on translated original sources, and wide reading in and skimming through the secondary literature. Ideally, such a guide as this should be wholly superfluous, since the historian's most faithful teacher is trial and error; but it may be welcomed by those whose time must be concentrated elsewhere, or who do not have browsing access to the stacks of one of the great Oriental collections. In some fields, such as the development of technology or the introduction of Western science, the important sources are too scattered to be brought together in a compilation as brief as this. They will be given extended consideration elsewhere.

This bibliography is critical in the sense that I have stressed the shortcomings of each work as well as its strong points, to the extent that there is point in doing so. It is necessary to remember that each book or article is important, or it would not have been included. Some works are important because they are highly useful or highly worth reading; this is not compromised in any way by their faults, which are stressed only so that they can be compensated for. Other works are important because they are widely cited or widely stocked, but are actually shoddy goods, and this needs to be pointed out. I trust that the reader will have no difficulty in distinguishing to which category I have assigned any given publication. A few particularly useful titles are marked with an asterisk, which signals a degree of enthusiasm that would be boring if expressed at greater length.

For the convenience of readers who must depend upon interlibrary loan arrangements, I have provided catalogue numbers for the Harvard-Yenching Library wherever possible. These are either unprefixed or prefixed by J, R, RW, or W. For other books, catalogue numbers are provided in the following order of preference and prefixed by the designation of the library: (1) Widener Library, Harvard University, (2) Humanities Library, M.I.T., and (3) other libraries.

Since the works cited use a variety of romanization conventions, the only common denominator of which is the presumably indecipherable ideogram, I have appended a conversion table that will allow reduction of all of them— assuming consistency on the part of the writers who use them, sometimes lamentably a mistaken assumption—to a single convention generally accepted by sinologists who write in English.

As with any tool, the real test of this bibliography is its utility. If those who receive it take issue freely with its form, its selections, and its judgments, the result will be a second edition of greater usefulness still. This version has greatly benefited from the thorough critique of P. van der Loon.

I. General Bibliography
*Cordier, Henri

Bibliotheca Sinica. Dictionnaire bibliographique des ouvrages relatifs à l'Empire chinois. 2d ed., 4 vols., Paris, 1904–1908. Supplement, Paris, 1924 (RW9540/14). Reprint, 5 vols., Taipei, 1966, also available.

The basic bibliography for Chinese studies to about 1920. An analytic list of books and periodical articles, exhaustive for post-1840 publications and valuable for earlier ones. See particularly the section "Sciences et arts." Works are arranged chronologically under headings. Extremely inconvenient classification. Supplemented by *Author Index to the Bibliotheca Sinica of Henri Cordier* [Second Edition...] (Mimeographed; New York: East Asiatic Library, Columbia University, 1953, RW9540/14i) which contains many errors but is indispensable.

Far Eastern Bibliography
Ithaca, New York, 1941–.
Continuation of the *Bulletin of Far Eastern Studies* (1936–1940), selection equally hit-or-miss, restricted to works in Western languages. Completely unannotated, so that its value is minimal. Published as part of *Far Eastern Quarterly* (later known as *Journal of Asian Studies*). Quarterly through Vol. 5, divided only into "Books" and "Articles." Thereafter annual and grouped by broad subjects.

Franke, Herbert
Sinologie. Bern, 1953 (RW9163/29).
A bibliographic essay in German covering the whole of classical Chinese studies, with rather more emphasis on cultural than political history. Far from exhaustive, but well chosen and well organized. Excellent for an idea of what has been done on a given question. For science and technology, see pp. 190–203.

*Hucker, Charles O.
China: A Critical Bibliography. Tucson, 1962 (RW9160/42).
The first place to go to find the best work on any sinological topic. "... A selected, graded, annotated list of books that contribute significantly, in the compiler's opinion, to the academic study of both traditional and modern China." Annotations, both descriptive and critical, are short. Unfortunately the compiler was able to spend very little space on science and technology (pp. 72–73).

Leslie, Donald; Jeremy **Davidson**
Author Catalogues of Western Sinologists (Guide to Bibliographies on China and the Far East). Canberra, 1966 (RW 9163/49).

Remarkably comprehensive. An excellent short list of general bibliographies appears on pp. 241–242. Perhaps the most useful single list of works by author is Charles S. **Gardner,** *Bibliographies of Sinologists* (Cambridge, 1958, RW 9696.6/31).

*Lust, John
Index Sinicus. A Catalogue of Articles Relating to China in Periodicals and Other Collective Publications, 1920–1955. Cambridge, England, 1964 (RW9540/52).

Supplements Cordier; covers those materials that **Yuan** does not. Well arranged, carefully compiled, but somewhat carelessly edited; by no means exhaustive.

*Revue bibliographique de sinologie 1955–
Paris, 1957– (RW9540/75).

Immensely valuable analytical annual bibliography of current sinological articles and books in humanities and social sciences published in Chinese, Japanese, and European languages. Wide coverage, but far from exhaustive. Abstracts, 40–250 words in length, mostly in English, and signed, are by experts in the respective fields. Section VII, "Histoire des sciences," concentrates on astronomy, medicine, and technology. Since it takes two to seven years to compile a volume, one must still depend on *Far Eastern Bibliography* for latest publications.

Skačkov, Petr Emel'janovič
Bibliografija Kitaja. Moscow: Izdatel'stvo Vostočnoj Literatur, 1960 (W9690/81).

According to P. van der Loon, practically exhaustive for publications in Russian.

Stucki, Curtis W.
American Doctoral Dissertations on Asia, 1933–1962. Including Appendix on Master's Theses at Cornell University. Ithaca: Southeast Asia Program, Department of Far Eastern Studies, Cornell University, 1963 (RW2406/82).

Many of these works, although unpublished, are available on microfilm from University Microfilms. To supplement for the earlier period, use *Columbia University Master's Essays and Doctoral Dissertations on Asia, 1875–1956* (Mimeographed; New York: East Asiatic Library, Columbia University, 1957, W9569/50)

*Yuan, T'ung-li
China in Western Literature. A Continuation of **Cordier's** Bibliotheca Sinica. New Haven, 1958 (RW9540/98).

Covers books and monographs (including some offprints and dissertations) in English, French, German, and Portuguese, 1921–1957; check "Addenda" for 1957 publications. The arrangement is more useful than **Cordier's,** but there are no annotations. Despite its size, not exhaustive, and there are many incorrect characters and romanizations; in most cases Wade-Giles romanization has not been given for titles and names that are romanized according to other systems in the works cited. Includes a name index, and a short index to romanized titles of Chinese books referred to. There is a section on "Natural Science," pp. 526–561.

———

A Guide to Doctoral Dissertations by Chinese Students in America 1905–1960. Washington, 1961 (RW9568/35).

———

"Doctoral Dissertations by Chinese Students in Great Britain and Northern Ireland, 1916–1961," *Chinese Culture,* 1963, *4.4.*

———

"A Guide to Doctoral Dissertations by Chinese Students in Continental Europe 1907–1962," ibid., 1964, *5.3:* 92–156, *5.4:* 81–149, *6.1:* 79–98.

Arranged by country with statistical breakdown at end.

II. Biography
Giles, Herbert A.
A Chinese Biographical Dictionary. London and Shanghai, 1898. Various reprints (RW2257.6/33).

The standard guide for an earlier generation, never adequate but not yet replaced in any European language. **Giles'** dates are often questionable, mistranslations are frequent, and legend is mixed with fact. Except for the Sung and Ch'ing periods, one is still forced to depend on more recent works that are not primarily biographical. A topical guide is Irvin V. **Gillis** and **Yü** Ping-yueh, *Supplementary Index to Giles' "Chinese Biographical Dictionary"* (Peking, 1936, RW2257.6/33.1) Some corrections have been published by E. von **Zach,** "Einige Verbesserungen zu Giles' Chinese Biographical Dictionary," *Asia Major,* 1926, *3:* 545–568; and Paul **Pelliot,** "A Propos du 'Chinese Biographical Dictionary' de M. H. Giles," ibid., 1927, *4:* 377–389; and "Les Yi nien lou," *T'oung Pao,* 1927, *25:* 65–81.

Hummel, Arthur W. (ed.)
Eminent Chinese of the Ch'ing Period (1644–1912). 2 vols. Washington, 1943–1944 (RW2259.8/40).

Possibly the most valuable single reference work on late Chinese history in any language. A series of 800 biographical sketches by experts, each with bibliography. End matter: (1) A list of subjects of the biographies in chronological order, with dates. (2) Index of all names mentioned (over 2000). (3) Index of books mentioned; many are pre-Ch'ing. (4) Exhaustive subject index. By use of this index the work may be used as a guide to general history of the period.

Weng, T'ung-wen
Répertoire des dates des hommes célèbres des Song (Matériaux pour le manuel de l'historie des Song, nr. IV). Paris, 1962 (RW2665/82.4).

For the period 960–1280. Accuracy high for a reference of this sort. See review by James T. C. **Liu** in *Journal of Asian Studies,* 1963, *22:* 323–324.

III. Translations
*Davidson, Martha
A List of Published Translations from Chinese into English,

French and German: Literature, Exclusive of Poetry. (Tentative edition.) Washington, 1952 (RW5044).

A comprehensive index of wide utility. There is another volume on poetry. For other translations, use **Yuan** T'ung-li, *China in Western Literature* (esp. pp. 61–65) and John **Lust,** *Index Sinicus* (esp. pp. 69–75).

Frankel, Hans H.
Catalogue of Translations from the Chinese Dynastic Histories for the Period 220–960. Berkeley, 1957 (RW2517/11.2).

Index to over 2000 passages translated from the third to the eighteenth of the Standard Histories. Parallel passages cross-referenced. Translator and subject indexes. The latter may be used to locate translations from the treatises on astronomy, and so on. No evaluations of translations, which vary in quality.

Needham, Joseph
Science and Civilisation in China. 7 vols. projected. Cambridge, England, 1954– (RW2470/59).

For specifically scientific matter, use of the indexes and bibliographies in Needham is the surest means of access.

IV. Background

A. HISTORIOGRAPHY

Beasley, W. G.; E. G. **Pulleyblank** (eds.)
Historians of China and Japan. (Historical Writing on the Peoples of Asia.) London, 1961 (W2460/07).

A collection of specialist articles of high quality, somewhat miscellaneous but all important.

*Gardner, Charles S.
Chinese Traditional Historiography. Cambridge, 1938. Reprint with additions and corrections by **Yang** Lien-sheng. Cambridge, 1961 (RW2460/31).

The standard general work in the field.

Ku Chieh-kang (trans. Arthur W. **Hummel**)

The Autobiography of a Chinese Historian. Leiden, 1931 (W2515/47.2). Reprint, Taipei[, ca. 1969].

Originally published 1926. **Ku** is a pioneer in the Chinese application of "scientific" scholarship to ancient history and presided at the unravelling of legend from history that shortened China's accepted record by two thousand years. Provides a glimpse into the mind of a traditionally trained but iconoclastic scholar of the first rank. **Hummel's** long introduction (pp. i–xlii) is an excellent discussion of the skeptical tradition in classical historiography.

Watson, Burton

Ssu-ma Ch'ien, Grand Historian of China. New York, 1961 (W2511/92).

On the life and work of the author of the *Shih chi,* the first of the series of Standard Histories. Extended treatment of ancient historiography, focussing on the greatest of China historians.

B. GENERAL HISTORY

Chang, Kwang-chih

The Archeology of Ancient China. Revised and enlarged ed., New Haven and London, 1968 (first publ. 1963; W2068/12.1).

A brilliant synthesis of recent archeological and literary evidence for the emergence and early history of Chinese civilization. Its interpretations supersede almost all of those available in historical surveys.

*De Bary, Wm. T.; **Chan** Wing-tsit; Burton **Watson**

Sources of Chinese Tradition. New York, 1960 (W2470/03). Also revised Columbia University Press paperback.

A survey of Chinese history, with emphasis on intellectual aspects, in the form of a large selection of short, usually abridged, translations. The introductions to each section are of high quality.

Fairbank, John K.; Edwin O. **Reischauer**
East Asia: The Great Tradition. (A History of East Asian Civilization, Volume I.) Boston, 1958, 1960 (W2420/75).

Written as a text for a beginning course in Far Eastern history, this is a masterly synthesis of the results of modern critical research. Unfortunately perfunctory on intellectual issues.

Franke, Otto
Geschichte des chinesischen Reiches. Eine Darstellung seiner Entstehung, seines Wesens und seiner Entwicklung bis zur neuesten Zeit. 5 vols. Berlin, 1930–1952 (W2510.29).

The most copious survey in a Western language. Vols. III and V are notes and indexes to the others. This work, unfinished at **Franke's** death (covers only to 1368), is primarily a political history, of considerable value as such but limited in other fields. Based mainly on Chinese sources. The prewar volumes are especially comprehensive. For a good review by O. van der **Sprenkel,** see *Bulletin of the School of Oriental and African Studies*, 1956, *18:* 312–321.

*Goodrich, L. Carrington
A Short History of the Chinese People. Rev. ed.; New York, 1951 (W2516/34). Also Harper Torchbooks paperback.

Short but quite reliable. Emphasizes the material aspects of civilization. Useful chronological chart on pp. 257–258. The best general history of its size for the historian of science.

Needham, Joseph
Science and Civilisation in China. Vol. I: Introductory Orientations. Cambridge, England, 1954 (RW2470/59).

Of very high quality are the "Bibliographical Notes" (pp. 42–54), the "Geographical Introduction" (pp. 55–72), the "Historical Introduction" (actually one of the best very concise surveys in print; pp. 73–149), and the long section on "Conditions of Travel of Scientific Ideas and Techniques between China and Europe" (pp. 150–248). The table,

"Transmission of Mechanical and Other Techniques from China to the West," on p. 242, appears in revised and expanded form in Charles **Singer,** et al. (eds.), *A History of Technology,* Vol. II (Oxford, 1956), pp. 770–771.

Wright, Arthur F.
"The Study of Chinese Civilization," *Journal of the History of Ideas,* 1960, *21:* 233–255.

"A general, interpretive review of the changing character of Western scholarship (and to some extent Chinese and Japanese scholarship) in reference to China during the 19th and 20th centuries."—C. O. **Hucker.**

C. ECONOMIC AND SOCIAL HISTORY

Bodde, Derk
"Feudalism in China," in Rushton **Coulborn** (ed.), Feudalism in History. Princeton, 1956, pp. 49–92 (W2320. 14).

For an understanding of the problem read the editor's introduction too. **Bodde** characterizes Chinese "bureaucratic feudalism" as economic rather than political. He uses a wider definition than would please most European historians, but his discussion remains the best place to begin. One should not fail, however, to proceed to two short critiques by J. R. **Levenson** (*Far Eastern Quarterly,* 1956, *15:* 369–372) and Etienne **Balazs** (*Journal of Asian Studies,* 1957, *16:* 329–332, reprinted in *Chinese Civilization and Bureaucracy* [New Haven, 1964], pp. 28–33), both of whom defend the more rigorous thesis that "China (200 B.C.–A.D. 1900) was not a feudal but a bureaucratic society. ...."

Sun, E-tu Zen; John **De Francis**
Chinese Social History. Washington, 1956 (W4130/83.1).

"Translations of 25 articles by modern Chinese scholars, chiefly dating from the 1930's, relating to socioeconomic aspects of Chinese life from earliest antiquity into the 19th century; an invaluable reference."—C. O. **Hucker.**

Sung Ying-hsing
T'ien-kung K'ai-wu. Chinese Technology in the Seventeenth Century. University Park, Pa., and London, 1966 (W8298/83).

Appendix C (pp. 362–363), "The Equivalence of Chinese Weights and Measures in Metric Units," arranged by dynasty, is based on Wu Ch'eng-lo's standard history of metrology in Chinese. For the history of length measures, **Wang** Kuo-wei, "Chinese Foot-measures of the Past Nineteen Centuries," *Journal of the North China Branch, Royal Asiatic Society*, 1928, *59*: 111–123, has never been superseded.

Swann, Nancy Lee
Food and Money in Ancient China. Princeton, 1950 (W2550/6883).

Standard economic history based on classical texts covering antiquity to the Han. See the substantial review by **Yang** Lien-sheng in *Harvard Journal of Asiatic Studies*, 1950, *13*: 524–557. Certain aspects, particularly government economic controls, have been carried to the seventh century in Rhea C. **Blue,** "The Argumentation of the *Shih-huo chih* Chapters of the Han, Wei, and Sui Dynastic Histories," ibid., 1948, *11*: 1–118.

Yang Lien-sheng
Money and Credit in China (Harvard-Yenching Institute Monograph Series XII). Cambridge, 1952 (W4352/98).

Technical and rather strictly devoted to the factual detail of the monetary and credit systems. Highly reliable.

D. PHILOSOPHY

Bibliographie Bouddhique
Paris, 1923–.
Standard fully annotated annual bibliography.

*Chan Wing-tsit
An Outline and an Annotated Bibliography of Chinese Philosophy. New Haven, 1961 (WR1010/4.13).

The first half is a detailed outline with relevant reading; bibliography in second half includes short critical notes. As authoritative as possible, almost exhaustive. Author partial to neo-Confucianism, to his own writings (justly), and to the ignorant effusions of F. S. C. **Northrop.** Includes articles.

––––––––

A Source Book in Chinese Philosophy. Princeton, 1963 (RW1011/12). Also Princeton U. Press paperback.

Excellent documentary survey, all selections freshly translated by **Chan.** Many were also used in *Sources of Chinese Tradition* (New York, 1960; full citation in Sec. IVB), which serves the same purpose less intensively.

Forke, Alfred

Geschichte der alten chinesischen Philosophie. Hamburg, 1927 (W1012/29). Geschichte der mittelalterlichen chinesischen Philosophie. Hamburg, 1934 (W1014/29). Geschichte der neueren chinesischen Philosophie. Hamburg, 1938 (W1020/29).

Not a coherent synthesis, but useful for translations from Chinese sources, and as a general orientation. The second volume in particular deals with thinkers ignored elsewhere. See E. von **Zach,** *Weitere Verbesserungen zu Forke's Geschichte der chinesischen Philosophie III. Bd.* (Offprint from *Mededeelingen van het China Instituut,* Batavia, July 1939, W1011/2998).

*Fung Yu-lan [**Feng** Yu-lan] (trans. Derk **Bodde**)

A History of Chinese Philosophy. Vol. I. Peking, 1937, rev. ed., Princeton, 1952. Vol. II. Princeton, 1953 (W1101/27).

The most reliable work in the field, critical and supported by lengthy and apt quotations. The Chinese original was published in 1934. It is advisable to check interpretations here (influenced by Western neorealism) with those in **Fung's** *A Short History of Chinese Philosophy* (ed. Derk **Bodde**; New York, 1948, paperback reprint, 1960; influenced by Taoist transcendentalism). But remember that this isolates only ideological, not historical, preconceptions; such key issues as the dates and relations of the *Lao-tzu* and the *Chuang-tzu* are still very much unsettled, and no single source can give a wholly balanced view. Good bibliography, which includes only works mentioned in the text.

Maspero, Henri
Mélanges posthumes sur les religions et l'histoire de la Chine. 3 vols. Paris, 1950 (W9163/53).

Vol. I ("Les religions chinoises," including a short essay on the introduction of Buddhism into China) and Vol. II ("Le Taoisme," equally concerned with the sect and the philosophy) are unexcelled

*Needham, Joseph
Science and Civilisation in China. Volume II: History of Scientific Thought. Cambridge, England, 1956 (RW2470/59).

Actually a survey of classical philosophy from the scientific point of view. Should be used with its point of view in mind: a combination of Whiteheadian organicism, socioeconomic determinism of a highly modified Marxian nature, elements of British empiricism, and a tendency to give China the benefit of the doubt in questions of priority (see the article of Shigeru **Nakayama** earlier in this volume). **Needham** is much more intellectually daring than **Fung** Yu-lan, and some of his interpretations have provoked a good deal of controversy. His presentation of the "philosophy of organism" has been generally accepted as a *tour de force* of comprehension, but other interpretations—e.g. the idea that the roots of the scientific and democratic tendencies in China are to be found in Taoism—have been sharply questioned. His theses are always fairly presented and fully documented, with alternate points of view noted. For good critiques, look up the references in **Hucker's** bibliography, p. 55.

Porter, Lucius Chapin
Aids to the Study of Chinese Philosophy. Peking, 1934 (RW1010.4/71).

There is very little of value in this book except for a series of comparative chronological charts on pp. 26–30.

Siu, Ralph G. H. [**Hsiao** Ken-k'ai]
The Tao of Science: An Essay on Western Knowledge and Eastern Wisdom. Cambridge, 1957 (Widener Ch404.58). Also M.I.T. Press paperback, 1964.
   Wu-wei for directors of American-style research establishments. Representative of a sizable genre of fuzzy-minded books that promise to apply the lessons of philosophical Taoism to self-improvement, but that succeed only in the measure that they avoid clear explications and concrete suggestions. Shows poor philosophical and historical understanding.

Welch, Holmes
The Parting of the Way: Lao Tzu and the Taoist Movement. Boston, 1957 (W1071/93). Also Beacon Press paperback.
   The most detailed discussion in English of the development of popular Taoism, based on **Maspero** and Chinese historians. Published in paperback as *Taoism: The Parting of the Way* (Boston, 1966). Should be supplemented by the works of such modern French students as Max **Kaltenmark,** Rolf **Stein,** Michel **Soymié,** and Kristofer **Schipper;** the most accessible and least specialized of these books is **Kaltenmark,** *Lao tseu et le taoïsme* (Maitres spirituels; Paris, 1965), translated into English, lamentably minus the remarkable illustrations, as *Lao Tzu and Taoism* (tr. Roger Greaves; Stanford, 1969, W1071/42). A digest of some important recent studies is provided in **Welch,** "The Bellagio Conference on Taoist Studies," *History of Religions,* 1969–1970, *9.* 2–3: 107–136.

V. History of Science

A. GENERAL

*Bodde, Derk
"Harmony and Conflict in Chinese Philosophy," in Arthur F. **Wright** (ed.), Studies in Chinese Thought. Chicago, 1953,

pp. 19–80 (W1010.9/95). Also University of Chicago Press paperback.

One of the most penetrating attempts to arrive at a synthesis of Chinese cosmology and values.

"Evidence for 'Laws of Nature' in Chinese Thought," *Harvard Journal of Asiatic Studies,* 1957, *20:* 709–727.

Issue published 1959. Takes issue with **Needham,** who doubts the word "law" is applicable to Chinese natural conceptions. Concludes, ". . . At least a few early Chinese thinkers viewed the universe in terms strikingly similar to those underlying the Western concept of 'laws of nature.' "

Chang Wing-tsit
"Neo-Confucianism and Chinese Scientific Thought," *Philosophy East and West,* 1957, *6:* 309–322.

Challenges **Needham's** overview of Chinese science. One of the more (but still not very) sophisticated attempts to explain the failure of China to develop modern science.

Japanese Studies in the History of Science
Annual. Tokyo, 1962–.

The only Western-language journal of which a major portion is devoted to Chinese science. Contributions are accepted from outside Japan. Vol. I contains a series of review articles.

*Needham, Joseph
Science and Civilisation in China. 7 vols. projected. Cambridge, England, 1954– (RW2470/59).

"The single work of scholarship which in our time has raised the banner of human unity most bravely and most triumphantly" (Philip **Morrison**). Based on a great mass of scattered research in European languages and work in the sources by the author and several Chinese collaborators. Audacious, literate, and arranged for maximal accessibility, with enormous bibliographies. **Needham** takes the worthwhile risk of forming a coherent interpretation of a field that is just being explored. The reader is advised to read critically and to check sources whenever possible, just as with any other pioneering work of synthesis.

———

The Grand Titration. Science and Society in East and West. London, 1969.

A preview of **Needham's** arguments (to be developed in Vol. VII) for the primacy of social and economic factors in the conditioning of Chinese scientific achievement. Particularly important is his survey of noncyclical time conceptions, "Time and Eastern Man," also published separately as the Henry Myers Lecture for 1964 (London, 1965) and in J. T. **Fraser** (ed.), *The Voices of Time* (New York, 1966), pp. 92–135. One can also follow in this collection of papers, published from 1944 on, the development and modification of the author's views. See the review in *Journal of Asian Studies,* 1971, *30:* 870–873.

*———

Clerks and Craftsmen in China and the West. Lectures and Addresses on the History of Science and Technology. Cambridge, England, 1970.

Another collection of contributions separately published from 1946 on, dealing with a spectrum of themes from the most general aspects of Asian scientific work seen in world perspective to articles on the earliest snow crystal observations and the early history of the rudder.

Sivin, Nathan
"Chinese Conceptions of Time," *The Earlham Review,* 1966, *1:* 82–92.

On the cosmological background of the early Chinese time sense.

Yabuuti, Kiyosi
"The Development of the Sciences in China from the 4th

to the End of the 12th Century," *Journal of World History*, 1958, *4:* 330–347.

Reprinted in *The Evolution of Science* (Mentor paperback, 1963), pp. 108–127. Execrably edited but occasionally useful chronological account of scientific and technological developments.

B. MATHEMATICS

Mikami, Yoshio

The Development of Mathematics in China and Japan. Leipzig, 1913; reprint, New York, n.d. (W7030/55).

Almost the only monograph of any value, but uncritical as to dates and in other respects. Although this book and A. **Wylie's** *Chinese Researches* (Shanghai, 1897, W9160/96, and reprints) are definitely still worth consulting (particularly for details of operations), they are partly superseded by J. **Needham,** *Science and Civilisation in China*, Vol. III. For readers who want to examine the primary literature, the only integral translation of a mathematical treatise that can be recommended is Kurt **Vogel's** rendering of the *Chiu chang suan shu,* which was compiled at some time during the last two centuries B.C., *Neun Bücher arithmetischer Technik* (Ostwalds Klassiker der exakten Wissenschaften, n. s., Vol. IV; Braunschweig, 1968). Important forthcoming translations with introductory studies include **Lam** Lay Yong on **Yang** Hui (fl. 1261/1275) and Ulrich **Libbrecht** on **Ch'in** Chiushao (fl. 1247).

Struik, Dirk

"On Ancient Chinese Mathematics," *The Mathematics Teacher,* 1963, *56:* 424–432.

A concise orientation, with excellent examples, based on Russian sources as well as **Mikami** and **Needham.**

C. ASTRONOMY

Chatley, Herbert

"Ancient Chinese Astronomy," *Occasional Notes of the Royal Astronomical Society,* 1939, no. 5, pp. 65–74.

A short résumé of the characteristics of Chinese astronomy.

Eberhard, Wolfram
"Index zu den Arbeiten über Astronomie, Astrologie und Elementenlehre," *Monumenta Serica,* 1942, *7:* 242–266.

An alphabetical index to ten important articles on Chinese astronomy, astrology, and cosmology. Two, one on the later Han calendar and one a translation of an important calendrical-cosmological treatise of the mid-second century, are in English. Another important article is too recent to be covered: "The Political Function of Astronomy and Astronomers in Han China," in John K. **Fairbank** (ed.), *Chinese Thought and Institutions* (Chicago, 1957, W2517/26), pp. 33–70. Before consulting the latter it is helpful to read the critique of **Eberhard's** approach to the history of science on pp. 53–54 of the article by **Sivin** listed in this section. **Eberhard's** astronomical essays are now being reprinted in book form by the Chinese Materials and Research Aids Service Center, Taipei, under the title *Sternkunde und Weltbild im alten China: Gesammelte Aufsätze.*

Ginzel, F. K.
Handbuch der mathematischen und technischen Chronologie. Das Zeitrechnungswesen der Völker. 3 vols. Leipzig, 1906–1914.

An introduction to the basic problems of calendar and ephemerides construction and their solutions in various cultures; none of the available literature on Chinese astronomy provides full introductory information of this kind. For a much more compact collection of data, see *Explanatory Supplement to the Astronomical Ephemeris and the American Ephemeris and Nautical Almanac* (London, 1961), pp. 407–433.

Ho Peng Yoke [Ping-yü]
The Astronomical Chapters of the Chin shu. With Amendments, Full Translation and Annotations. Paris, 1966 (W7108/39).

This is a definitive translation of the Astrological Treatise (*T'ien wen chih*), which provides abundant data on positional astronomy and the astrological interpretation of observational data—not on mathematical astronomy, which is discussed in the Treatise on Harmonics and the Calendar (*Lü li chih*). Ho supplements this first translation of its kind with full explanatory notes and valuable introductory remarks on historiography and observatory practice.

*Maspero, Henri
"L'astronomie chinoise avant les Han," *T'oung Pao*, 1929, *26:* 267–356.

"Les instruments astronomiques des Chinois au temps des Han," *Mélanges chinois et Bouddhiques*, 1938–1939, *6:* 183–370.
Rigorous, accurate, and illuminating. Basic for early astronomy, and among **Needham's** main sources.

Nakayama, Shigeru
"Characteristics of Chinese Astrology," *Isis*, 1966, *57:* 442–454.
A concise characterization, with new information on the late importation of horoscopic astrology of the Hellenistic type.

*Needham, Joseph
Science and Civilisation in China. Vol. III: Mathematics and the Sciences of the Heavens and the Earth. Cambridge, England, 1959, pp. 169–641 (RW2470/59).
Excellent survey, with much attention to observational astronomy and instruments. Most of the detailed research in the field has been done by Japanese astronomers, but it has not been used here or elsewhere in Western languages. See, for instance, the very concise review by Shigeru **Nakayama,** "Japanese Studies in the History of Astronomy," *Japanese Studies in the History of Science*, 1962, *1:* 14–22. In view of the fact that Chinese astronomy was almost entirely calendrical (that is, oriented toward the production of a complete ephemerides), the reader is misled when **Needham** dismisses the calendrical problem as such as trivial, and deals with the

various aspects of astronomy as though they were autonomous. Generally highly accurate; in case of doubt, check sources.

*Schlegel, Gustave

Uranographie chinoise, ou preuves directes que l'Astronomie est originaire de la Chine et qu'elle a été empruntée par les anciens peuples occidentaux à la sphère chinoise. 2 vols. with star maps in folder. La Haye, Leyde, 1875 (W7140/79). Reprint, Taipei, 1967.

Schlegel's extravagant views on the antiquity of Chinese astronomy in no way vitiate the eminence of this work as the basic reference on Chinese constellations and stars. A more complete list of stars, unfortunately useless to those who lack Chinese, is Alexander **Wylie,** "List of Fixed Stars," in *Chinese Researches,* pp. 346ff.

Sivin, Nathan

"Cosmos and Computation in Early Chinese Mathematical Astronomy," *T'oung Pao,* 1969, *55:* 1–73. Also published separately, Leiden, 1969.

An attempt to trace the changing roles of metaphysical assumptions in forming the mathematical techniques of early astronomy

Yabuuti, Kiyosi

"Astronomical Tables in China, from the Han to the T'ang Dynasties," in **Yabuuti** (ed.), *Chūgoku chūsei gijutsushi no kenkyū* (Studies in the history of medieval Chinese science and technology). Tokyo, 1963, pp. 445–492 (J7016/4430.1).

A systematic survey of the Chinese calendrical treatises and the astronomy they reflect, technically sophisticated. Continued in "Astronomical Tables in China from the Wutai to the Ch'ing Dynasties," *Japanese Studies in the History of Science,* 1963, *2:* 94–100, a tabular résumé without discussion.

D. ALCHEMY AND EARLY CHEMISTRY

Chikashige, Masumi

Alchemy and other Chemical Achievements of the Ancient

Orient: The Civilization of Japan and China in Early Times as Seen from the Chemical Point of View. Tokyo, 1936 (W7308/12).

The ambitious title aside, the first half of this book is an early speculative attempt, restricted to the *Pao p'u tzu nei p'ien*, at making chemical sense of ancient alchemy. The rest is an interesting but not always sound discussion of ancient bronze metallurgy and the famous Japanese art of sword making. Romanization is erratic.

Davis, Tenney L.
"The Dualistic Cosmogony of Huai-nan-tzu and its Relations to the Background of Chinese and of European Alchemy," *Isis,* 1936, *25:* 327–340.

Eliade, Mircea
The Forge and the Crucible. New York, 1962.

Original title *Forgerons et alchimistes* (Paris, 1956). A distinguished general study of the primitive roots of alchemical doctrines. The chapters on Chinese and Indian alchemy are of high quality. For a survey of recent developments, see "The Forge and the Crucible: A Postscript," *History of Religions,* 1968, *8:* 74–88.

Ho Ping-yü; Joseph **Needham**
"The Laboratory Equipment of the Early Medieval Chinese Alchemists," *Ambix,* 1959, *7:* 57–115.

Suggests a line of evolution parallel to but different from that of the European still. Valuable for its short history of the Taoist Patrology, in which the alchemical classics are preserved.

Leicester, Henry M.; Herbert S. **Klickstein**
"Tenney Lombard Davis and the History of Chemistry," *Chymia,* 1950, *5:* 1–16.

Davis, a professor of chemistry at M.I.T., edited and published a number of translations of classical alchemical works done by his graduate students, and wrote several seminal studies in the field. An extensive bibliography of these writings is on pp. 6–10. While the translations vary greatly in quality, only that of the *Pao p'u tzu nei p'ien* (ca. A.D. 320) has been bettered in part. See James R. **Ware** (ed. and tr.), *Alchemy, Medicine, Religion in the China of A. D. 320. The Nei P'ien of Ko Hung (Pao-pu-tzu)* (Cambridge, Mass., 1966, W1187/92). **Ware's** equation of *Tao* with God, explained on pp. 1–2, is not accepted by most sinologists.

Li Ch'iao-P'ing
The Chemical Arts of Old China. Easton, Pa., 1948 (W7308/ 50).
   Little space is devoted to alchemy and little is said about it that is not said better elsewhere. This is, however, an excellent informal compendium of Chinese chemical technology, explaining the traditional methods of making everything from alloys to soy sauce and wine. For translated materials on seventeenth-century chemical technology, see the book of **Sung** Ying-hsing cited in Sec. IVC.

Sivin, Nathan
Chinese Alchemy: Preliminary Studies (Harvard Monographs in the History of Science, Vol. I). Cambridge, Mass., 1968 (W7308/81).
   Mostly discussions and heuristic studies aimed at developing a critical and analytical approach to the history of alchemy. Among the appendixes is a list of published translations of alchemical treatises.

––––––––

"Chinese Alchemy as a Science," *Transactions of the International Conference of Orientalists in Japan,* 1968, *13:* 117–129.
   A general sketch of the relation of theory and practice in alchemy, and of some of the characteristics that define its role in the history of science. Based on more recent research than the preceding item.

Waley, Arthur
"Notes on Chinese Alchemy," *Bulletin of the School of Oriental and African Studies,* University of London, 1930, *6:* 1–24.

An erudite and still useful attempt to get at the beginnings of Chinese alchemy and to assess the possibility of foreign influence. The only flaw of any consequence is **Waley's** suggestion (due to ignorance of technical language) that the earliest extant treatise, the *Chou i ts'an t'ung ch'i,* is not at all concerned with chemical operations. An equally influential but very poorly informed attack on the same problems has been made by Homer H. **Dubs**; the latest version is "The Origin of Alchemy," *Ambix,* 1961, *9:* 23–36.

Wilson, William Jerome
"Alchemy in China," *CIBA Symposia,* 1940, *2:* 594–624 (Offprint, W7308/94).

Includes a comprehensive annotated bibliography to date; the article itself is of little value.

E. MEDICINE

Chamfrault, A.; **Ung** Kan Sam
Traité de médecine chinoise. Vol. I. Acupuncture. Moxas. Massages. Saignées. Angoulême, France, 1954. Vol. II. Les livres sacrés de médecine chinoise. 1957. Vol. III. Pharmacopée. 1959. Vol. IV. Formules magistrales. 1961. Vol. V. De l'astronomie à la médecine chinoise: Le ciel. La terre. L'homme. 1963 (W7910/87).

A mass of translated and paraphrased material of fair quality. Vol. II is a paraphrase of the *Huang ti nei ching* (The inner [= esoteric?] classic of the Yellow Emperor, ca. first century A.D.), the supreme canonic work of Chinese medicine. A more "literal" partial translation, Ilza **Veith** (trans.), *Huan Ti Nei Ching Su Wen, The Yellow Emperor's Classic of Internal Medicine* (Baltimore, 1949; W7910/36) is much inferior; see the review of the 1966 reprint in *Isis,* 1968, *59:* 229–231. Vol. V, done by **Chamfrault** alone, suffers from lack of access to Chinese sources.

Chang, Lucy Gi Ding
"Acupuncture: A Selected Bibliography," *Chinese Culture,* 1963, *5:* 156–160.

An unannotated list of 46 books in French, German, and Italian, mostly published between 1940 and 1960. Only a small fraction will hold interest for anyone except the practitioner or amateur of acupuncture as practiced in Europe today. This bibliography supplements and extends the list of books and articles in **Ferreyrolles.**

Chen Chan-yuen (**Ch'en** Ts'un-jen)
Chung-kuo i-hsueh-shih t'u chien ("History of Chinese medical science illustrated with pictures"). Hong Kong, 1968.

As the first pictorial archive of the subject this book is invaluable, particularly since condensed English versions of the captions are provided. But nothing in the sloppy text should be used without checking; the author has missed no opportunity to represent legends as historical facts and late imaginative depictions as portraits. The final pages, labeled "Spread of Chinese Medical Science into Great Britain," France, the United States, Japan, etc., contain mostly snapshots of the author's travels.

Chinese Journal of Medical History
Shanghai, 1947–1948. (National Library of Medicine).

The first Western-language journal devoted to the history of science published in China, unfortunately abortive. *The Chinese Medical Journal* (W7901/12) has since published an excellent series on the "Achievements of Chinese Medicine ..." by **Lee** T'ao.

Demiéville, Paul
"Byō," in **Demiéville** (ed.), Hōbōgirin. Dictionnaire encyclopédique du Bouddhisme d'après les sources chinoises et japonaises. Paris, 1937, pp. 224–265.

A magisterial survey of the connections between sickness, medicine, and Buddhism—medicine as a metaphor for

Buddhism, illnesses of the Buddha and Boddhisatvas, medicine in monastic discipline and in Buddhist charity, Buddhist theories of sickness, and the relations of the religion to the Indian and Sino-Japanese medical traditions. Offprints of this article were issued with the subtitle "Maladie et médecine dans les textes bouddhiques."

Ferreyrolles, Paul
L'acupuncture chinoise. Lille, 1953 (Library of Congress RM184-F4).
Recommended solely for the bibliography of European writing on Chinese medicine from 1671 to 1950 (pp. 177–191).

Huard, Pierre
"Quelques aspects de la doctrine classique de la médecine chinoise," *Biologie médicale,* 1957, *46:* i–cxix (N. Y. Academy of Medicine).
A systematic survey of physiology and pathology, the most useful work by this author. Additional information on anatomy and physical anthropology can be found in E. T. **Hsieh,** "A Review of Ancient Chinese Anatomy," *Anatomical Records,* 1921, *20:* 97–127, but it is extremely unreliable for names and dates.

————; Ming **Wong** (**Wang** Ming)
"Bio-bibliographie de la médecine chinoise," *Bulletin de la Société des études indo-chinoises,* n. s., 1956, *31:* 181–246 (Widener Ind 19.6[31. 3]).
Index of principal Chinese medical writers, and brief bibliography of Western (including Russian) and Chinese works.

————
"Évolution de la matière médicale chinoise," *Janus,* 1958, *47:* 3–67.
A conspectus of pharmacological literature, well illustrated but not highly reliable. Useful for a general impression of the tradition but does not render **Bretschneider,** "Botanicon sinicum," *16:* 22–95, obsolete (see below, Sec. VF)

La médecine chinoise au cours des siècles. Paris, 1959.

The best general history available, but pedestrian and not highly accurate or critical. Half the book is chronological and the other half topical.

Chinese Medicine (tr. Bernard **Fielding;** World University Library). New York, Toronto, 1968.

This beautifully illustrated original paperback deserves praise for its concern with the connections of other Asiatic traditions with that of China, with European knowledge of Chinese medicine and vice versa in the last three centuries, and with modern and traditional medicine in contemporary China. At the same time, it is so perfunctory, so carelessly thought through, and so full of elementary errors of fact, interpretation, translation, and transliteration that it should not be cited for data that cannot be verified elsewhere.

Ghani, A. R.

Chinese Medicine and Indigenous Medical Plants (Pakistan National Scientific and Technical Documentation Center, PANSDOC Bibliography No. 396). Karachi, 1965.

An unannotated list of about 300 books and papers in European languages and Chinese, included because of availability to the compiler rather than coverage or value. Most are either nontechnical or concerned with individual plants. The majority of important Chinese publications in medicine and pharmacology since 1949 are omitted. Almost all titles are given in English.

Hübotter, Franz

Die chinesische Medizin zu Beginn des XX. Jahrhunderts und ihr historischer Entwicklungsgang. Mimeographed. Leipzig, 1929 (W7908.40).

Not a history, but a rather miscellaneous collection of translated extracts and illustrations. The short historical sketch at the end does not distinguish events of different periods. There is a 43-page atlas of Chinese anatomy. See the extended review by George **Sarton** in *Isis,* 1930, *14:* 255–261.

Hume, Edward H.
The Chinese Way in Medicine. Baltimore, 1940.
A mélange of irrelevant anecdote and inaccurate information.

Lu Gwei-djen; Joseph **Needham**
"Records of Diseases in Ancient China," in Don **Brothwell** and A. T. **Sandison** (eds.), Diseases in Antiquity. A Survey of the Diseases, Injuries, and Surgery of Early Populations. Springfield, Ill., 1967 (M.I.T. Humanities R135.B874).
Invaluable survey of disease entities known to the Chinese through the first millennium B.C., with attention to their conceptualization. Many will find the attempt to show that leprosy is mentioned in the Confucian Analects unconvincing.

Mann, Felix
Acupuncture. The Ancient Art of Healing. New York, 1963.
By the President of Britain's Medical Acupuncture Society, this book reflects a high-minded ignorance of Chinese history, philosophy, science, and medical thought. It is, however, less hysterical than other books of its kind and thus is useful as a description of modern European acupuncture.

*Needham, Joseph; **Lu** Gwei-djen
"Medicine and Chinese Culture," in **Needham,** Clerks and Craftsmen in China and the West (Cambridge, England, 1970), pp. 263–293.
By far the best discussion to date of the social background of medicine, although some interpretations are forced (see above, p. 270). This volume contains three other articles of only slightly less general interest: "Proto-endocrinology in Medieval China" (pp. 294–315), "Hygiene and Preventive Medicine in Ancient China" (pp. 340–378), and "China and the Origin of Qualifying Examinations in Medicine" (pp. 379–395).

Read, Bernard E.
Chinese Materia Medica. Peking, 1931–1941 (RW7970/75.8).
Standard references for animal drugs. For detailed list of fascicules, see J. **Needham,** *Science and Civilisation in China* (Cambridge, England, 1959), III, 784–785.

————; C. **Pak**
A Compendium of Minerals and Stones Used in Chinese Medicine From the Pen Ts'ao Kang Mu... [of] Li Shih Chen... 1597 A.D. 2d ed., Peking, 1936 (RW7970/75.8).
Unlike *Chinese Materia Medica,* this is not a translation but a somewhat ill-digested mass of information from hither and yon. The order of accuracy is, however, high.

Roi, Jacques
Traité des plantes médicinales chinoises (Encyclopédie biologique, Nr. 47). Paris, 1955.
The most detailed and reliable work on medical plants in a Western language. An index to earlier treatises of the same sort is Bernard E. **Read,** *Chinese Medicinal Plants from the Pen Ts'ao Kang Mu...* A.D. 1596, 3rd ed. (Peking, 1936). There are indexes to plant names in Latin and French, but not in Chinese. Romanization is nonstandard.

Smith, F. Porter
Chinese Materia Medica. Vegetable Kingdom. Revised edition, ed. G. A. **Stuart.** Shanghai, 1911. Reprint (ed. Ph. Daven **Wei**), Taipei, 1969.
The fullest of the older monographs on Chinese materia medica. Still worth consulting, particularly in view of its indexes to Chinese names of herbs, which **Roi** lacks. The reader should also not overlook the translated material on identities and characteristics of medicinal plants in **Bretschneider,** "Botanicon sinicum," Part III (see Sec. V.F below).

Soulié de Morant, Georges
L'acuponcture chinoise... La tradition chinoise classifiée,

precisée. Paris, 1957 (Columbia University Medical Library).

This edition brings together two earlier volumes (Paris, 1939, 1941) and three posthumous volumes by the enthusiast and practitioner most responsible for the vogue of acupuncture in Europe. His works are based on a narrow selection of late compilations and provide an unsatisfactory picture of classical rational medicine.

Wong, K. Chimin [**Wang** Chi-min]; **Wu** Lien-teh
History of Chinese Medicine. Being a Chronicle of Medical Happenings in China from Ancient Times to the Present Period. 2d ed., Shanghai, 1932 (W7908/95).

Only about a quarter is devoted to traditional medicine and is hopeless as history. The rest is about the impact of European medicine; since it ignores Chinese sources, it is almost entirely an institutional history. This book's materials on the Chinese tradition have been treated in more synoptic and critical fashion by Willy **Hartner,** "Heilkunde im alten China," *Sinica,* 1941, *16:* 217–265; 1942, *17:* 266–328. This work is extremely rare, almost all copies having been destroyed by bombs during World War II, but a photographic copy of an offprint is on deposit in the Harvard-Yenching Institute.

Wu Wei-ping
Chinese Acupuncture. Translated and adapted from the Chinese, with added comments, by J. **Lavier**. English edition translated and adapted by Philip M. **Chancellor.** Rustington, Sussex [1962?] (National Library of Medicine WB369. W955c. 1964).

This is perhaps the first book in a Western language by a classically trained practitioner. Unfortunately most of it is the usual list of points.

F. BIOLOGY

Note. In the classical period biology did not exist as a distinct discipline. The bulk of Chinese biological knowledge is preserved in the great pharmacopoeias (since practically every substance was used in one or another function in Chinese prescriptions) or in agricultural works; with some exceptions, Chinese biological thought is inextricable from general philosophy. There is, therefore, little to recommend in this particular section. Standard reference works on zoology, botany, and so on, which do not pay attention to history, are omitted, since they are fully listed in Tung-li **Yuan,** *China in Western Literature,* pp. 549–561 (see p. 284 in this volume).

Bretschneider, E.
"Botanicon sinicum. Notes on Chinese Botany from Native and Western Sources," *Journal of the North China Branch, Royal Asiatic Society,* 1881–1895, Vols. 16 (Bibliographical sources), 25 (Botanical identifications), 29 (Uses in medicine). Fascimile reprints, Tokyo, 1937; Nendeln, Lichtenstein, 1967 (W3079/9).

"The Study and Value of Chinese Botanical Works," *Chinese Recorder,* 1870, *3:* 157–163, and in following issues.
    The author, a famous physician and botanist, studied most of the Chinese writings specifically devoted to botany. The second work has been paid the compliment of translation into Chinese.

Liou-ho [Liu Hou]; Claudius **Roux**
Aperçu bibliographique sur les anciens traités chinois de Botanique, d'Agriculture, de Sériciculture et de Fungiculture. Lyon, 1927 (Library of Congress Z5358.C517).
    A charming but unscholarly run-through, more up-to-date than **Bretschneider's** survey but otherwise inferior.

Needham, Joseph
"The Development of Botanical Taxonomy in Chinese Culture," in Actes du douzième congrès international d'histoire des sciences. Paris, 1968, pp. 127–133.
    Tentative but suggestive.

Nguyen Tran Huan
"Esquisse d'une histoire de la biologie chinoise des origines

jusqu'au IVe siècle," *Revue d'histoire des sciences,* 1957, *10:* 1–37.

Summary account of early developments.

A Conversion Table for Seven Systems of Romanization

The problems of developing an internationally satisfactory system of transcription for Chinese characters and gaining international acceptance for it are a favorite source of controversy among dyspeptic sinologists. The pressure of publishers limits the choice of symbols to the letters of the standard alphabet and thus rules out the only ultimately sensible solution, adoption of the International Phonetic Alphabet. Agreement on any new system would not affect the vast bulk of sinological literature already written. It has come to be generally accepted that a more pressing problem is to unite the sinological world, generally divided into the Wade-Giles and French areas (the Russian and Japanese systems pose less tractable and less urgent problems). The German system, which formerly was an additional obstacle, has largely been replaced by Wade-Giles. No system in widespread use has more than negligible merits as (1) indicating to the sinologist the meaning of the characters transcribed (which is impossible, since tones are not indicated; and many syllables, for example, *yu,* represent dozens of common characters), (2) providing the untrained reader with even a fair approximation to the Chinese pronunciation (for example, the "u" in *yu* and the "ou" in *kou* stand for the same sound; the "u" in *yu* and the "u" in *ku* stand for different sounds). The objections to unification are thus parallel to those for the unification of Europe, and the ink still spills.

In order to form a conception of the dimensions of the problem, consider the book title *T'ung-chih lüeh* (Wade-Giles), which would appear as follows in other systems: Legge, *T'ung-chih lio;* Wylie, *T'ung-che lĕŏ;* French, *T'oung-tche lio;* German, *Tung-dschi lüä;* Karlgren, *T'ung-tsi lüe;* and Needham, *Thung-chih lüeh.*

The point for the student of Chinese science is that he must adopt a single system for his own writings. Since the Wade-Giles system is standard in the English-speaking world, it is recommended. The personal system of Joseph Needham, like others of its kind, incorporates no substantial practical improvement, and has created only confusion outside his own circle of collaborators. The Pinyin system deserves special consideration because of its adoption by the Chinese government, but so far it has not been used to any extent, even in China, in scholarly writings about ancient Chinese science. A table follows for converting transcriptions in other systems the user of this bibliography is likely to come across.

In recent years many sinologists in the United Stated have made a practice of dropping diacritical marks that perform no function (the distinction between "yu" and "yü" is functional, since they represent syllables of different pronunciation; but the diacritic in "yüeh" serves no purpose; there is no corresponding "yueh" of different pronunciation). Finals that include such optional diacritics are marked by the symbol = in the table.

In order to save space, the roughly four hundred possible Chinese syllables are represented by a system of twenty-one initials, thirty-nine finals, and four syllables best treated separately. The division in every case falls before the first vowel. Finals that may stand alone as syllables are marked with a plus sign (+); special forms in other romanizations that hold only when the final stands alone as a syllable are enclosed in parentheses. Two or more entries separated by a slash have the same Wade-Giles equivalent. The reader is warned that this table is a reliable guide only for books or articles whose authors have *mastered* the system of romanization they have adopted, and are wholly consistent. In actuality, transliterations are often slapdash; this is especially true, for obvious reasons, of Chinese or Japanese scholars writing in Western languages. The only solution may be to prevail upon an Orientalist to check your transcriptions.

For access to other systems, the most complete handbook available is:

Legeza, Ireneus Laszlo
Guide to Transliterated Chinese in the Modern Peking
Dialect. 3 vols. projected. Leiden, 1968–.

For references to writings on problems of romanization, see:

Yang, Winston L. Y.; Teresa S. **Yang**
A Bibliography of the Chinese Language. New York, 1966,
pp. 70–76,

in addition to appropriate sections of general bibliographies.

Comparison of Romanizations

W-G	Legge	Wylie	French	German	Karlgren	Needham
Initials						
*ch-	ch-/k-	k-/ch-/ts-	k-/ts-/tch-	dj-/dsch-	k-/ts-	
ch'-		k-/ch-/ts-	k'-/ts'-tch'-	tj-/tsch-	k'-/ts'-	chh-
f-						
h-						
hs-		h-/s-	h-/s-	h-	h-/s-	
j-	j-/z-					
*k-				g-		
k'-				k-		kh-
l-						
m-						
n-						
*p-				b-		
p'-				p-		ph-
s-						
sh-			ch-	sch-		†
*t-				d-		
t'-				t-		th-
*ts-				ds-		
ts'-	ts'-/sh-			ts-		·tsh-
w-						
y-			i-(omitted before i)			
Finals						
+-a	-â					
+-ai	-âi	-aé (gae)				
+-an	-ân	/ên (gan)	/(ngan)			

Table, continued

W-G	Legge	Wylie	French	German	Karlgren	Needham
+-ang	-ăng		/(ngang)			
+-ao	-âo	-aou/oò/ew	/(ngac)	-au		
= +-ê		-îh	-o	-o/ö		
-ei		-eê	/(ngo)	-e		
-en	-ăn/en	-ùn		-ën		
= +-ên	-ăn/un	-în	/(ngen)	-ën		
= -êng	-ăng	-ăng/îng/ung	-ong	-ëng		
+-i/(yi/i)	-î/(yi/i)	-è/-eih(yih/e)				
-ia	-iâ	-ëá				
-iai	-ieh	-eaè				
-iang		-ëáng				
-iao	-iâo	-eaou		-iau		
-ieh		-ëĕ	-ie/-iai	-iä	-ie	
-ien		-eĕn		-iän		
-ih		-e	-e	-ï	-ï	
-in						
-ing						
= -io/üeh		-ëŏ	-io/iue	-üä		
-iu	-iû	-ew	-ieou			
-iung		-eung	-iong			
+-o		/(go)				
+-ou	-âu	-ow (gow)	-eou			
-u	-û	-oo/uh/ŏ	-ou			
-ü		-eu/ŭh				
-ua	-wâ	-wa	-iua			
-uai	-wäi	-wae	-ouai			
-uan	-wan	-wan	-ouan			
-üan		-uên/euèn	-iuan			
-uang	-wang	-wang	-ouang			
= -üeh	-io	-io/-iue/èŏ	-io/-iue	-üä	-üe	
-uei	-wei	-wuy/wei	-ouei	-ue		
-ui	-ûi	-uy	-ouei	-ui		
-un			-ouen			
-ün	-eun	-eun	-iun			
-ung			-ong			
-uo	-wô	-wo	-ouo			

Syllables

+erh	r	ùrh	eul	erh	er	
+jih	zah	jeh/jih	je	jï	jï	
+tzû	tsze	tsze	tseu	dsï	tsï	

Table, continued

W-G	Legge	Wylie	French	German	Karlgren	Needham
+tz'û	ts'ze	ts'ze	ts'eu	tsï	ts'ï	tzhu
+ssu	sze	sze	sseu	sï	sï	

*Unaspirate—not pronounced as read.
†Gives "sêng" for "sheng."

W-G: Wade-Giles, now used by writers in most languages except French. Division into initials and finals is modified from that of Joseph Needham, *Science and Civilisation in China* (Cambridge, England, 1954), I, 24–25.

Legge: The system used by James Legge in his standard translations of the Confucian classics. Reconstructed from *The Chinese Classics* (second ed., Oxford, 1893), *passim*. This is not the same system as that used in his other translations, now seldom consulted, in the Wisdom of the East Series.

Wylie: Used by Alexander Wylie, who wrote several works of value to the history of science. Reconstructed from *Chinese Researches* (Shanghai, 1897) and *Notes on Chinese Literature* ("new edition," Shanghai, 1902), *passim*. Wylie's romanization, especially his use of diacritical marks, is not highly consistent.

French: The system used by *Bulletin de l'École française d'Extrême-Orient*, and by almost all French writers. Compiled from syllable-by-syllable chart in James R. **Ware**, *Vocabularies to the Intermediate Chinese Texts at Harvard University* (Cambridge, 1937). See also the recent *Tables des concordances pour l'alphabet phonetique chinois* (Maison des sciences de l'homme. Materiaux pour l'étude de l'Extrême-orient moderne et contemporaine: études linguistiques, 2; La Haye, 1967).

German: Used by most prewar German writers. Taken from the "Tafel der Lautumschreibung" (Lessing-Othmar system) in Werner Rüdenberg, *Chinesisch-Deutsches Wörterbuch* (Hamburg, 1936), which gives Wade-Giles equivalents.

Needham: His modification consists primarily of replacing the aspiration diacritic (a superscript comma) by the letter "h"; Joseph Needham, *Science and Civilisation*, I, 24–25.

Karlgren: Another personal system, designed for (but not much used by) philologists. Bernhard Karlgren, *The Romanization of Chinese* (London, 1928), 10–14.

Index